Inverse and Ill-Posed Problems Series

Ill-Posed and Non-Classical Problems of Mathematical Physics and Analysis

Also available in the Inverse and Ill-Posed Problems Series:

Forward and Inverse Problems for Hyperbolic, Elliptic and Mixed Type Equations
A.G. Megrabov
Nonclassical Linear Volterra Equations of the First Kind
A.S. Apartsyn
Poorly Visible Media in X-ray Tomography
D.S. Anikonov, V.G. Nazarov, and I.V. Prokhorov
Dynamical Inverse Problems of Distributed Systems
V.I. Maksimov
Theory of Linear Ill-Posed Problems and its Applications
V.K. Ivanov, V.V. Vasin and V.P. Tanana
Ill-Posed Internal Boundary Value Problems for the Biharmonic Equation
M.A. Atakhodzhaev
Investigation Methods for Inverse Problems
V.G. Romanov
Operator Theory. Nonclassical Problems
S.G. Pyatkov
Inverse Problems for Partial Differential Equations
Yu. Ya. Belov
Method of Spectral Mappings in the Inverse Problem Theory
V. Yurko
Theory of Linear Optimization
I.I. Eremin
Integral Geometry and Inverse Problems for Kinetic Equations
A.Kh. Amirov
Computer Modelling in Tomography and Ill-Posed Problems
M.M. Lavrent'ev, S.M. Zerkal and O.E. Trofimov
An Introduction to Identification Problems via Functional Analysis
A. Lorenzi
Coefficient Inverse Problems for Parabolic Type Equations and Their Application
P.G. Danilaev
Inverse Problems for Kinetic and Other Evolution Equations
Yu.E. Anikonov
Inverse Problems of Wave Processes
A.S. Blagoveshchenskii
Uniqueness Problems for Degenerating Equations and Nonclassical Problems
S.P. Shishatskii, A. Asanov and E.R. Atamanov
Uniqueness Questions in Reconstruction of Multidimensional Tomography-Type Projection Data
V.P. Golubyatnikov
Monte Carlo Method for Solving Inverse Problems of Radiation Transfer
V.S. Antyufeev
Introduction to the Theory of Inverse Problems
A.L. Bukhgeim
Identification Problems of Wave Phenomena - Theory and Numerics
S.I. Kabanikhin and A. Lorenzi
Inverse Problems of Electromagnetic Geophysical Fields
P.S. Martyshko
Composite Type Equations and Inverse Problems
A.I. Kozhanov
Inverse Problems of Vibrational Spectroscopy
A.G. Yagola, I.V. Kochikov, G.M. Kuramshina and Yu.A. Pentin
Elements of the Theory of Inverse Problems
A.M. Denisov
Volterra Equations and Inverse Problems
A.L. Bughgeim
Small Parameter Method in Multidimensional Inverse Problems
A.S. Barashkov
Regularization, Uniqueness and Existence of Volterra Equations of the First Kind
A. Asanov
Methods for Solution of Nonlinear Operator Equations
V.P. Tanana
Inverse and Ill-Posed Sources Problems
Yu.E. Anikonov, B.A. Bubnov and G.N. Erokhin
Methods for Solving Operator Equations
V.P. Tanana
Nonclassical and Inverse Problems for Pseudoparabolic Equations
A. Asanov and E.R. Atamanov
Formulas in Inverse and Ill-Posed Problems
Yu.E. Anikonov
Inverse Logarithmic Potential Problem
V.G. Cherednichenko
Multidimensional Inverse and Ill-Posed Problems for Differential Equations
Yu.E. Anikonov
Ill-Posed Problems with A Priori Information
V.V. Vasin and A.L. Ageev
Integral Geometry of Tensor Fields
V.A. Sharafutdinov
Inverse Problems for Maxwell's Equations
V.G. Romanov and S.I. Kabanikhin

INVERSE AND ILL-POSED PROBLEMS SERIES

Ill-Posed and Non-Classical Problems of Mathematical Physics and Analysis

PROCEEDINGS OF THE INTERNATIONAL CONFERENCE
SAMARKAND, UZBEKISTAN

Editor-in-Chief:
M.M. Lavrent'ev
Managing Editor:
S.I. Kabanikhin
Editorial Board:
Akb.H. Begmatov, T.D. Djuraev,
S. Saitoh and M. Yamamoto

///VSP///
UTRECHT • BOSTON
2003

VSP
P.O. Box 346
3700 AH Zeist
The Netherlands

Tel: +31 30 692 5790
Fax: +31 30 693 2081
vsppub@compuserve.com
www.vsppub.com

First published in 2003

ISBN 90-6764-380-7

Printed in The Netherlands by Ridderprint bv, Ridderkerk.

Invited papers

Contributed papers

This book is the Proceedings of the International Conference "Ill-Posed and Non-Classical Problems of Mathematical Physics and Analysis" which was held at the Samarkand State University, Samarkand, Uzbekistan from 11 to 15 September 2000.

The Conference was organized jointly by the Samarkand State University and Sobolev Institute of Mathematics, Novosibirsk, Russia. Advice and general guidance were provided by the International Programme Comittee. More than 90 participants from Germany, Japan, Kazakhstan, Russia, Tajikistan, Turkey, and Uzbekistan presented their lectures at the Conference.

The scientific program of the Conference covered the following topics:

- Theory of Ill-Posed Problems
- Inverse Problems for Differential Equations
- Boundary Value Problems for Equations of Mixed Type
- Integral Geometry
- Mathematical Modelling and Numerical Methods in Natural Sciences

The Proceedings bring together fundamental research articles in the major areas of the numerated fields of analysis and mathematical physics. The papers in this volume represent all plenary and some contributed lectures presented at the conference. All the papers have undergone peer review.

We would like to thank the Conference paricipants for their interesting reports and stimulating discussions. We express our sincere thanks to the authors for submitting articles of such high quality and to the referees for their thoughtful reviews. Finally, we would like to thank the staff at VSP for their help in publishing the Proceedings.

M. M. Lavrent'ev
S. I. Kabanikhin
Akb. H. Begmatov
T. D. Djuraev
S. Saitoh
M. Yamamoto

Ill-Posed and Non-Classical Problems of
Mathematical Physics and Analysis, Samarkand, 2000, pp. 1–18
M. M. Lavrent'ev and S. I. Kabanikhin (Eds)

Problems of integral geometry on curves and surfaces in Euclidean space

Akbar H. BEGMATOV* and Akram H. BEGMATOV†

1. INTRODUCTION

Integral geometry studies the transformations assigning to functions on a manifold X their integrals over submanifolds from certain set M [23]. This important and intensively developing domain of modern mathematics is closely connected with the theory of partial differential equations, mathematical physics, geometric analysis. This direction has many applications in mathematical study of problems of seismic exploration, interpretation of the data of geophysical and aerospace observations, in solving inverse problems of astrophysics and hydroacoustics (see [33] and the references given there). Methods developed here are basic for solving the problems of medical and industrial tomography [17, 18, 34].

The problem of recovery of a function from its integrals over all possible hyperplanes in Euclidean space was considered by J. Radon [37]. In this classical work the explicit inversion formulas for even and odd-dimensional spaces were obtained. Also, the methods of solution of such problems were developed and the various analogs of this integral transformation were considered [37]. Note that P. Funk [21], the year before, had investigated the problem of recovery of even function in the sphere from its the integralsover big circles.

*Novosibirsk, Sobolev Institute of Mathematics. E-mail: bah@math.nsc.ru
†Samarkand State University, Uzbekistan. E-mail: akrbegmat@mail.ru

The work was partially supported by State Commitee of Science and Technology of Uzbek Republic (grants Nos. 15/99 and 2/01).

The connection of problems of integral geometry with differential equations was investigated by J. John [26, 27] (see, also, [28]). The transformation (called after by the ray transformation) which assigns to a function in three three-dimensional space its integrals over all possible straight lines was considered in [27].

Basing on the theory of Lie groups, I. M. Gel'fand with co-authors had considered the problems of integral geometry on linear manifolds. This had become the basic for investigation of integral geometry for wide classes of homogeneous spaces. Interesting inversion formulas were obtained here. The problems of reconstructing a function from its integrals over surfaces of second order were also widely considered (see [33] and the bibliography cited here).

The first important results for a multidimensional inverse kinematic seismic problem were obtained by M. M. Lavrent'ev and V. G. Romanov [32] by the method of its reducing to the problem of determining the function from its mean values along all possible circles with centers in a fixed line. This work brought the attention to the cases when integration goes along the manifolds of more complicated geometric structure. Namely, in [32] the connection between multidimensional inverse problems for partial differential equations and the problems of integral geometry was revealed. The manifolds which arise after reducing the inverse problems to the problems of integral geometry are naturally connected with the initial differential equation. For the equation with variable coefficients these geometric objects may be sufficiently complicated.

V. G. Romanov had investigated the problems of uniqueness and stability in the case when the manifolds are invariant under the group of all motions parallel to a $(n-1)$-dimensional hyperplane [38, 39]. The weight functions are also assumed to be invariant under this group.

The important results on uniqueness and stability of solution were obtained by Yu. E. Anikonov and A. L. Bukhgeim for the following classes of integral geometry problems: the solution is sought in the classes of functions analytic with respect to some variables; the manifolds along which the integrations goes depend analytically on some variables (parameters) (see [1, 16]).

Questions of uniqueness, obtaining stability estimates and inversion formulas for various classes of problems of integral geometry in Euclidean space were investigated in [2–14].

In the second section of our paper uniqueness questions for a wide class of problems of integral geometry in the plane are considered. Unlike the works mentioned above we do not impose the conditions of invariance or analyticity. The uniqueness theorem is established for smooth convex curves.

The third section is devoted to weakly ill-posed problems of integral geometry on curves and surfaces with singularities in the vertices. The existence and uniqueness theorems and the explicit inversion formulas are obtained as well as the stability estimates in Sobolev spaces. For the problem of integral geometry on the family of cone surfaces in the even-dimensional space very simple representation of solution was obtained. On the basis of these results the stability estimates in Sobolev spaces were obtained; therefore, weak ill-posedness of the problem was established. The existence theorem was obtained also.

Note that in our case (the cone surfaces) the odd case and the even case differ essentially. Such situation occurs frequently in the problems of integral geometry.

Finally, in the fourth section we consider the problem of recovery of the function given the integrals of this function on a family of lines in three-dimensional space which are generatrices of cones. Such integral transformations are called ray transformations [25] and they have wide applications in computer tomography [34].

The inversion formulas of the ray transformation connected with the cone scheme of scanning the computer tomography may be found in the works of H. K. Tuy [40], D. V. Finch [20], P. Grangeat [24] (see, also, the surveys of F. Natterer [35], V. P. Palamodov [36] and the references cited there). A. A. Kirillov [29], A. S. Blagoveshchenskii [15], I. M. Gel'fand and A. B. Goncharov [22] are investigated various settings of the problems of recovery of a function from its integrals taken along the lines intersecting a certain set in the space. The explicit formulas of effective determining the desired function were obtained.

The problem that we consider is connected with auxiliary problems of analytic extension. Unlike the problems considered in [11–13] this problem is strongly ill-posed. The uniqueness theorem for solution of the problem in the class of continuous finite functions is established. The estimate of conditional stability of solution of the logarithmic type problem was obtained.

2. THE PROBLEM OF INTEGRAL GEOMETRY OF THE VOLTERRA TYPE OF GENERAL FORM IN THE PLANE

Let $\xi = (\xi_1, \xi_2)$, $y = (y_1, y_2)$, $x = (x_1, x_2)$, $\mathbf{R}^2_+ = \{x = (x_1, x_2) : x_2 \geq 0\}$. We consider the family of the curves $\{\Gamma(x)\}$ in the strip $L = \{x \in \mathbf{R}^2_+ : 0 \leq x_2 \leq h,\ h < \infty\}$. The curve from this set is defined by the equation $\xi_2 = \psi(x, \xi_1)$. We assume also that for any $x \in L$ and any angle $\alpha \in [0, \alpha_1] \cup [\alpha_2, \pi]$, where

$0 < \alpha_1 < \pi/2 < \alpha_2 < \pi$ there exists a unique curve from $\{\Gamma(x)\}$ passing through the point x at the angle α with positive direction of the axis OX_1.

We consider the problem of recovery of a function of two variables $u(x)$ if we know the integrals of this function with a given weight function $p(x, \xi_1)$ on the family $\{\Gamma(x)\}$. This is the problem of solution of the operator equation in the function $u(x)$:

$$\int_{\Gamma(x)} p(x, \xi_1) u(\xi)\, d\xi_1 = f(x). \tag{1}$$

We can represent equation (1) as follows:

$$\int_{x_1 - h_1}^{x_1 + h_2} p(x, \xi_1) u(\xi)\, d\xi_1 = f(x), \tag{2}$$

where $h_1 = h_1(x)$, $h_2 = h_2(x)$.

Theorem 1. *Let the right-hand side $f(x)$ of the equation (2) be known for all x from the strip L; $h_k \in \mathbf{C^2}$, $k = 1, 2$, $p \in \mathbf{C^3}$, $\psi \in \mathbf{C^7}$ and the following conditions hold*

$$p(x, \xi_1) \geq \delta_1 > 0; \quad \psi(x, x_1) = x_2;$$

$$\psi(x, \xi_1 - h_1) = \psi(x, \xi_1 + h_2), \quad \psi(x, x_1 - h_1) = \psi(x, x_1 + h_2) = 0;$$

$$-\infty < \delta_3 \leq \frac{\partial^2 \psi}{\partial \xi_1^2} \leq \delta_2 < -2,$$

where $\delta_1, \delta_2, \delta_3 (\delta_3 < \delta_2)$ are constants.

Then the solution of equation (2) is unique in the class $\mathbf{C}_0^2(L)$.

Proof. Pass in (2) to integration with respect to ξ_2

$$\int_0^{x_2} g(x, \xi_2)[u(x_1 - \varphi(x, \xi_2)) + u(x_1 + \varphi(x, \xi_2))]\, d\xi_2 = f(x).$$

Fix a certain point $x_0 \in L$. Taking into account the conditions imposed on the curves of the family $\{\Gamma(x)\}$, we can select the subset $\Gamma \subset \{\Gamma(x)\}$ of the curves being tangent from inside to the parabola $P(x^0) = \{x \in L : x_2 = x_2^0 - (x_1 - x_1^0)^2\}$. Let $t = (t_1, t_2)$ be the point of touching the curves $\Gamma(x)$ and $P(x^0)$. Then

$$t_1 = \beta_1(x^0, x), \qquad t_2 = \beta_2(x^0, x) \tag{3}$$

where β_1, β_2 are smooth functions and

$$t_2 = x_2^0 - (x_1^0 - t_1)^2, \qquad t_2 = \psi(x^0, t_1). \tag{4}$$

Formulas (3), (4) and the inverse function theorem yield

$$x_1 = \varphi_1(x^0, t_1) \qquad x_2 = \varphi_2(x^0, t_1),$$

where φ_1, φ_2 are smooth functions.

As a result we obtain $\xi_2 = \psi(x_1, x_2, \xi_1) = \psi(\varphi_1(x^0, t_1), \varphi_2(x^0, t_1), \xi_1) = \varphi(x^0, t_1, \xi_1)$, where φ is a smooth function. Thus, we pass from the parametrization of the curve of the family $\Gamma(x)$ using the coordinates of its vertex x_1, x_2 to the parametrization by means of the three parameters x_1^0, x_2^0, t_1. Analogously, $g(x, \xi_1) = g(\varphi_1, \varphi_2, \xi_1) = g_1(x^0, t_1, \xi_1)$, where g_1 is a smooth function.

We have

$$\int_0^{x_2} g(x, \xi_2)[u(x_1 - \varphi(x, \xi_2), \xi_2) + u(x_1 + \varphi(x, \xi_2), \xi_2)]\, d\xi_2$$
$$= \int_0^{\varphi_2(x^0, t_1)} [g_1(x, \xi_2) u(\varphi_1 - \psi_1, \xi_2) + g_2(x, \xi_2) u(\varphi_1 + \psi_2, \xi_2)]\, d\xi_2,$$

where φ_1, φ_2, ψ_1 and ψ_2 are smooth functions.

Thus, the initial problem is reduced to the following problem of integral geometry

$$\int_0^{\varphi_2(x^0, t_1)} [g_1(x, \xi_2) u(\varphi_1 - \psi_1, \xi_2) + g_2(x, \xi_2) u(\varphi_1 + \psi_2, \xi_2)]\, d\xi_2 = f(x^0, t_1). \tag{5}$$

on the family of curves $\{\Gamma(x^0, t_1)\}$. It is easy to verify that the right-hand side of equation (5) satisfies certain differential equation of the first order with partial derivatives; therefore, the problem of its solution is not overdetermined.

We apply now to both sides of (5) the integral operator

$$Ih(x^0, t_1) = \int_{x_1^0 - \sqrt{x_2^0}}^{x_1^0 + \sqrt{x_2^0}} K(x^0, t_1) h(x^0, t_1)\, dt_1.$$

As a result we obtain

$$If_1(x^0, t_1) = \int_{x_1^0 - \sqrt{x_2^0}}^{x_1^0 + \sqrt{x_2^0}} K(x^0, t_1) f_1(x^0, t_1)\, dt_1$$
$$= \int_{x_1^0 - \sqrt{x_2^0}}^{x_1^0 + \sqrt{x_2^0}} w(x^0, t_1) \int_0^{\varphi_2(x^0, t_1)} [g_1(x, \xi_2) u(\varphi_1 - h_1, \xi_2)$$
$$+ g_2(x, \xi_2) u(\varphi_1 + h_1, \xi_2)]\, d\xi_2\, dt_1$$

$$= \int_{x_1^0-\sqrt{x_2^0}}^{x_1^0+\sqrt{x_2^0}} \int_0^{\varphi_2(x^0,t_1)} \big[w(x^0,t_1)g_1(x,\xi_2)u(\varphi_1 - h_1,\xi_2)$$

$$+ w(x^0,t_1)g_2(x,\xi_2)u(\varphi_1 + h_1,\xi_2)\big]\, d\xi_2\, dt_1$$

$$= \iint_{P(x^0)} W(x^0,\xi)u(\xi_1,\xi_2)\, d\xi_1\, d\xi_2.$$

Here $P(x^0)$ is a part of the strip L bounded by OX_1-axis and the parabola $\mathcal{P}(x^0)$ with the vertex in the point x^0:

$$\mathcal{P}(x^0) = \{(\xi_1,\xi_2) : x_2^0 - \xi_2 = (x_1^0 - \xi_1)^2\}.$$

We choose the kernel K of the integral operator I so that the weight function $W(x^0,\xi) = W_0(x^0,\xi)/\sqrt{x_2^0 - \xi_2 - (x_1^0 - \xi_1)^2}$, where $W_0(x^0,\xi)$ is smooth function.

Then, applying the integro-differential operators, we can reduce the problem of solution of the latter equation (therefore, the initial problem of integral geometry) to the problem of integral geometry in the strip in the set of parabolas with the perturbation:

$$\int_{\mathcal{P}(x^0)} u(\xi_1,\xi_2)\, d\xi_1 + \iint_{P(x^0)} G(x^0,\xi)u(\xi_1,\xi_2)\, d\xi_1\, d\xi_2 = F(x^0). \tag{6}$$

The uniqueness of the solution of such problem of integral geometry with perturbation was investigated by M. M. Lavrent'ev [31]. The results of this paper imply that if the conditions of Theorem 1 hold, then the solution of equation (6) is unique in the class $\mathbf{C}_0^2(L)$. Hence Theorem 1 follows. □

3. WEAKLY ILL-POSED PROBLEMS: UNIQUENESS, THE STABILITY ESTIMATES AND THE INVERSION FORMULAS

Denote by Ω the strip

$$\Omega = \{(x,y) : x \in \mathbf{R}^1,\ y \in (0,l),\ l < \infty\}.$$

We consider the operator equations is a function $u(x,y) \in C_0^5(\Omega)$:

$$\int_0^y \big[u(x+h,\eta) + u(x-h,\eta)\big]\, d\eta = f_0(x,y) \tag{7}$$

and

$$\int_0^y \big[u(x+h,\eta)+u(x-h,\eta)\big]\,d\eta \\ + \int_0^y \int_{x-h}^{x+h} K(x,y,\xi,\eta)u(\xi,\eta)\,d\xi\,d\eta = f(x,y), \quad (8)$$

where $h=(y-\eta)^2$, $(x,y)\in\Omega$. The function $K(\cdot)\in C_0^5(\Omega\times\Omega)$ and vanishes on the curves $\mathcal{P}(x,y)=\{(\xi,\eta):(y-\eta)^2=|x-\xi|,\ 0\le\eta\le y,\ 0\le y\le l\}$ together with its derivatives up to the fifth order inclusively.

Suppose $\lambda\in\mathbf{R}^1$, $p\in\mathbf{C}$ ($p=p_1+ip_2$). Introduce the functions

$$\mathcal{J}(\lambda,p)=\int_0^\infty e^{-p\zeta}\cos(\lambda\zeta^2)\,d\zeta,$$

$$\mathcal{I}_1(\lambda,y)=\frac{1}{2\pi i}\int_{a-i\infty}^{a+i\infty} e^{py}\frac{dp}{\mathcal{J}(\lambda,p)p^3}, \qquad a>0,$$

$$\mathcal{I}_2(x,y)=\int_{-\infty}^{\infty} e^{i\lambda x}\frac{\mathcal{I}_1(\lambda,y)}{1+\lambda^2}\,d\lambda,$$

and denote

$$\partial_y^3 f_0(x,y)=\mathcal{F}_0(x,y)=\frac{\partial^3 f_0(x,y)}{\partial y^3}, \qquad d_x^2\mathcal{F}_0(x,y)=\Big(E-\frac{\partial^2}{\partial x^2}\Big)\mathcal{F}_0(x,y),$$

where E is the unit operator.

Theorem 2. *The solution of equation* (7) *in the class* $_0^5(\Omega)$ *is unique and it has the following representation*

$$u(x,y)=\int_0^y\int_{-\infty}^{\infty}\mathcal{I}_2(x-\xi,y-\eta)(d_x^2\partial_y^3 f_0)(\xi,\eta)\,d\xi\,d\eta,$$

and satisfies the inequality

$$\|u\|_{L_2(\Omega)}\le C_0\|f_0\|_{W_2^1(\Omega)}.$$

Here C_0 *is some constant;* $W_2^1(\Omega)$ *is the Sobolev space.*

In the proof of uniqueness of the solution of equation (7) we shall use the integral

$$\mathcal{J}(\lambda,p)=\int_0^\infty e^{-p\zeta}\cos(\lambda\zeta^2)\,d\zeta.$$

Making the change $\sqrt{|\lambda|}\zeta = \tau$, $\rho = \frac{p}{l}\sqrt{|\lambda|}$, we obtain $\mathcal{J}(\lambda, p) = \mathcal{J}_1(\rho)/\sqrt{|\lambda|}$, where

$$\mathcal{J}_1(\rho) = \int_0^\infty e^{-\rho\tau}\cos(\tau^2)\,d\tau, \qquad \rho \in \mathbf{C}\ (\rho = x + iy).$$

It occurs that the integral $\mathcal{J}_1(\rho)$ satisfies the estimates obtained below in Lemmas 1–3 (the proof of the lemmas can be found in [13]).

Lemma 1. *For $\mathcal{J}_1(x)$ the following estimate holds*

$$\mathcal{J}_1(x) > \begin{cases} 1/(4\sqrt{2\pi}\,), & \text{if } x \in [0, \sqrt{2/\pi}\,), \\ 1/(4\pi x), & \text{if } x \in [\sqrt{2/\pi}, \infty). \end{cases}$$

Lemma 2. *For each $x \geq 0$, $\mathcal{J}_1(x)$ satisfies the inequalities*

$$\mathcal{J}_1(x) \leq \frac{1}{2}\sqrt{\frac{\pi}{2}}, \qquad \cos\Big(\frac{x^2}{4} + \frac{\pi}{4}\Big)\mathcal{J}_1(x) \leq \frac{\sqrt{\pi}}{4}.$$

Moreover, the integral $\mathcal{J}_1(\rho)$ is separated from zero in the right half-plane.

Lemma 3. *Let $\rho \in \mathbf{C}_+ = \{\rho = x + iy : x > 0,\ y \in \mathbf{R}\}$. Then the integral*

$$\mathcal{J}_1(\rho) = \int_0^\infty e^{-\rho\tau}\cos\tau^2\,d\tau$$

satisfies the inequality

$$|\mathcal{J}_1(\rho)| > 0.$$

Proof of Theorem 2. At first we suppose that $l = \infty$. Applying the Fourier tranformation with respect to x and the Laplace transformation with respect to y to both sides of (7), we obtain

$$v(\lambda, p)\mathcal{J}(\lambda, p) = \psi(\lambda, p). \tag{9}$$

Here $v(\lambda, p)$ and $\psi(\lambda, p)$ are the Fourier—Laplace images of the desired function and the right-hand side of equation (7) respectively, and

$$\mathcal{J}(\lambda, p) = \int_0^\infty e^{-p\zeta}\cos(\lambda\zeta^2)\,d\zeta.$$

We try to obtain now the estimates from below of the function $\mathcal{J}(\lambda, p)$. This will allow us to obtain the estimates for $v(\lambda, p)$ and, finally, the estimates for the desired function $u(x, y)$. Taking into account Lemmas 1–3, we obtain

$$1/|\mathcal{J}| < 4\pi(\sqrt{|\lambda|} + |p|). \tag{10}$$

Apply to equation (9) the inverse Laplace transformation with respect to p and the Fourier transformation with respect to λ. Taking into account the inversion theorem and the convolution theorem and using the properties of these transformations, we obtain

$$u(x,y) = \int_0^y \int_{-\infty}^{\infty} \mathcal{I}_2(x-\xi, y-\eta)(d_x^2 \partial_y^3 f_0)(\xi,\eta)\, d\xi\, d\eta, \tag{11}$$

where $\mathcal{I}_2(x-\xi, y-\eta)$ and $(d_x^2 \partial_y^3 f_0)(\xi,\eta)$ are defined above.

The inversion formula (11) has the local character with respect to y. Taking into account the condition $\operatorname{supp} u \subset \Omega$, we have that representation (11) holds also for an arbitrary $l < \infty$. Then it follows from (9)–(11) that the solution of the initial problem of integral geometry (1) is unique in the class $C_0^5(\Omega)$, and the following estimate holds:

$$\int_{-\infty}^{\infty} \int_{a-i\infty}^{a+i\infty} |v(\lambda,p)|^2\, dp\, d\lambda \\ \leq \int_{-\infty}^{\infty} \int_{a-i\infty}^{a+i\infty} |4\pi(1+|\lambda|+|p|)|^2 |\psi(\lambda,p)|^2\, dp\, d\lambda.$$

This inequality and (11) yield the stability estimate for solution of equation (7). □

Theorem 3. *The solution of equation (8) is unique in the class $C_0^5(\Omega)$ and satisfies the following estimate*

$$\|u\|_{L_2(\Omega)} \leq C\|f\|_{W_2^1(\Omega)},$$

where C is a constant.

Proof. Consider the second term on the left-hand side of equation (8):

$$\int_0^y \int_{x-h}^{x+h} K(x,y,\xi,\eta)u(\xi,\eta)\, d\xi\, d\eta = f_1(x,y), \quad h = (y-\eta)^2.$$

For $y < y_0$, y_0 being sufficiently small, and taking into account the conditions that the weight function $K(\cdot)$ satisfies, we obtain the estimate

$$\|f_1\|_{W_2^1(\Omega)} \leq \varepsilon\|u\|_{L_2(\Omega)}, \tag{12}$$

where $\varepsilon \to 0$ as $y_0 \to 0$.

The estimates from Theorem 2 and the triangle inequality for the norms yield

$$\|u\|_{L_2(\Omega)} \leq C_0(\|f\|_{W_2^1(\Omega)} + \|f_1\|_{W_2^1(\Omega)}). \tag{13}$$

Inequalities (12) and (13) yield the inequality

$$\|u\|_{L_2(\Omega)} \le C\|f\|_{W_2^1(\Omega)}, \tag{14}$$

where

$$C = C_0/(1 - C_0\varepsilon).$$

Hence the existence of solution of equation (8) follows for y sufficiently small. Since (8) is a Volterra equation [33], the uniqueness and stability estimate (14) will hold not only for small y, but in the whole strip Ω. The theorem is proved. □

Introduce the notations $x = (x_1, x_2, \dots, x_{n-1})$, $\xi = (\xi_1, \xi_2, \dots, \xi_{n-1})$, $n > 2$; $y \in \mathbf{R}^1$, $\eta \in \mathbf{R}^1$, $y \ge 0$, $\eta \ge 0$, $\Omega = \{(x,y) : x \in \mathbf{R}^{n-1}, y \in (0,h), h < \infty\}$, $\bar{\Omega} = \{(x,y) : x \in \mathbf{R}^{n-1}, y \in [0,h]\}$.

In the layer $\bar{\Omega}$ consider the set of cones $\{K(x,y)\}$ with the vertices in the points (x,y)

$$K(x,y) = \Big\{(\xi,\eta) : \sum_{m=1}^{n-1}(x_m - \xi_m)^2 = (y-\eta)^2,\ 0 \le \eta \le y,\ y \le h,\ h < \infty\Big\}.$$

We denote by $Q(x,y)$ the part of n-dimensional space bounded by the surface $K(x,y)$ and the hyperplane $y = 0$. Consider the following problem

Find the function $u(\cdot)$ dependent on n variables if for all $(x,y) \in \bar{\Omega}$ the integrals of this function are known along the surfaces $K(x,y)$:

$$\iint_{K(x,y)} u(\xi,\eta)\,dk = f(x,y). \tag{15}$$

Theorem 4. *Suppose n is even and the function $f(\cdot)$ is known in the layer Ω. Then the solution of (15) in the class $\mathbf{C}_0^n(\Omega)$ is unique, the following representation*

$$u(x,y) = C_0\Big(\frac{\partial^2}{\partial y^2} - \Delta_x\Big)^{n/2}\int_0^y f(x,\eta)\,d\eta, \tag{16}$$

and the inequality

$$\|u(x,y)\|_{W_2^{0,\dots,0,1}(\Omega)} \le C_1\|f(x,y)\|_{W_2^{n,\dots,n}(\Omega)}.$$

hold. Here C_1 is a constant,

$$C_0 = \frac{1}{2^{n-1/2}\pi^{(n-1)/2}\Gamma(n/2)}, \qquad \Delta_x = \frac{\partial^2}{\partial x_1^2} + \dots + \frac{\partial^2}{\partial x_{n-1}^2}.$$

The theorem can be proved analogously Theorem 2.

The problem of existence of solution of equation (15) is formulated as follows.

Theorem 5. *Let the right-hand side of equation* (15) *satisfy the following conditions:*

1) $f(x,y)$ *is finite with respect to* x*;*

2) $f(x,y)$ *has all continuous partial derivatives up to* n*-th order;*

3) $\frac{\partial^k}{\partial y^k} f(x,y)\Big|_{y=0} = \frac{\partial^k}{\partial y^k} f(x,y)\Big|_{y=h} = 0 \quad (0 \le k \le n).$

Then the solution of equation (15) *exists in the class of continuous functions finite with respect to* x *and is defined by formula (16).*

4. INVERSION OF THE X-RAY TRANSFORMATION WITH INCOMPLETE DATA

Let $x = (x_1, x_2, x_3) \in \mathbf{R}^3$, $\xi = (\xi_1, \xi_2, \xi_3) \in \mathbf{R}^3$, $\lambda = (\lambda_1, \lambda_2, \lambda_3) \in \mathbf{R}^3$, $\bar{x} = (x_1, x_2) \in \mathbf{R}^2$, $\bar{\xi} = (\xi_1, \xi_2) \in \mathbf{R}^2$, $\bar{\lambda} = (\lambda_1, \lambda_2) \in \mathbf{R}^2$, $Q = \mathbf{R}^3 \times \Theta$, $\mathcal{D} = \mathbf{R}^2 \times \Theta$, $\Theta = \{\alpha : \alpha \in [0, 2\pi]\}$, $\mu = (\cos\alpha, \sin\alpha)$, $\Omega = \{(x,y) : |x| < 1, |y| < l, 0 < l < \infty\}$; $\bar{\Omega} = \{(x,y) : |x| \le 1,\ |y| \le l\Big\}$.

Denote by $K(x)$ the family of two-sheeted cones with vertices in the points (x):

$$K(x) = \{\xi \in \mathbf{R}^3 : |x_3 - \xi_3| = |\bar{x} - \bar{\xi}|\}.$$

We consider the operator equation in a function $u(x)$

$$\int_{\mathbf{R}^1} u(x_1 + s\cos\alpha, x_2 + s\sin\alpha, x_3 + s)\, ds = f(x, \alpha). \tag{17}$$

We assume that the function $f(\cdot)$ is known for all $(x, \alpha) \in Q$.

Thus, we consider the problem of integral geometry for the family of lines which are generatrices of two-sheeted cones $K(x)$. The desired function $u(\cdot)$ is the function of three variables and the right-hand side of equation (17) depends on four parameters x_1, x_2, x_3, and α. Nevertheless, the problem of solution of equation (17) is not overdetermined since its right-hand side satisfies the equation with partial derivatives of the first order

$$\frac{\partial f}{\partial x_1}\cos\alpha + \frac{\partial f}{\partial x_2}\sin\alpha + \frac{\partial f}{\partial x_3} = 0.$$

Moreover, the following theorem on uniqueness of solution of equation (17) holds.

Theorem 6. *Solution of equation* (17) *is unique in the class of continuous finite functions with the support in* Ω.

Proof. Applying to both sides of equation (17) the Fourier transformation with respect to the variables x_1, x_2, we obtain

$$\int_{\mathbf{R}^1} e^{-is\langle\bar{\lambda},\mu\rangle} w(\lambda_1, \lambda_2, s + x_3)\, ds = \psi(\bar{\lambda}, x_3, \alpha).$$

We denote here by $w(\bar{\lambda}, x_3)$ and $\psi(\bar{\lambda}, x_3, \alpha)$ the Fourier transformations with respect to x_1, x_2 of the functions $u(x)$ and $f(x, \alpha)$ respectively, and $\langle\cdot\,,\cdot\rangle$ is the scalar product.

Changing the variables $s+x_3 = t$ and introducing the notation $\lambda_1 \cos\alpha + \lambda_2 \sin\alpha = -\lambda_3$ we obtain

$$v(\lambda) = e^{ix_3\lambda_3}\psi(\lambda, \alpha), \tag{18}$$

where $v(\lambda)$ is the Fourier transformation of the function $w(\lambda_1, \lambda_2, x_3)$ with respect to x_3.

In the space of the Fourier variables $(\lambda_1, \lambda_2, \lambda_3)$ consider the two-sheeted cone

$$\mathcal{K} = \{(\lambda_1, \lambda_2, \lambda_3) \in \mathbf{R}^3 : |\bar{\lambda}| = |\lambda_3|\}$$

with the vertex in the origin.

Denote

$$\tilde{\mathcal{K}} = \{\lambda \in \mathbf{R}^3 : |\bar{\lambda}| < |\lambda_3|\}.$$

Taking into account that (18) holds for all $\alpha \in [0; 2\pi]$ and $\psi(\bar{\lambda}, x_3, \alpha)$ is the Fourier transformation with respect to the variables (x_1, x_2) of the given function $f(x, \alpha)$, we see that the values of $v(\lambda)$ are uniquely determined from equation (18) for all $\lambda \in \mathbf{R}^3 \setminus \tilde{\mathcal{K}}$. Thus, the problem of finding $u(x)$ from equation (17) is reduced to the problem of extension of its Fourier image $v(\lambda)$ from $\mathbf{R}^3 \setminus \tilde{\mathcal{K}}$ to the interior of the cone $\mathcal{K}$.

Since u is finite, $v(\lambda)$ can be extended analytically with respect to the first argument into the complex plane. Therefore, the initial problem of integral geometry is reduced to the following problem of analytical extension: to find the function $v(\nu_1 + i\nu_2)$ in the interval $\{\nu \in \mathbf{C} : |\nu|_1 < \lambda_3^*,\ \nu_2 = 0\}$ which is analytical in the whole complex plane if its values in the rays $L_1 = \{(\nu_1, \nu_2) : \nu_1 \geq \lambda_3^*,\ \nu_2 = 0\}$ and $L_2 = \{(\nu_1, \nu_2) : \nu_1 \leq -\lambda_3^*,\ \nu_2 = 0\}$ are known.

The solution of this problem, as is known, is unique. Hence Theorem 6 follows. □

We shall formulate now one auxiliary statement which will be useful for deduction of stability estimate for the problem in question.

In the complex plane $\mathbf{C}$ we consider the strip

$$S = \{z = z_1 + iz_2 : z_1 \in \mathbf{R}^1, \ |z_2| < a\pi, \ a > 0\}$$

and the rays

$$r_1 = \{z : -\infty < z_1 \leq -a, \ z_2 = 0\}, \quad r_2 = \{z : a \leq x < \infty, \ y = 0\}.$$

Suppose that $G = S \setminus \{r_1 \cup r_2\}$. In other words, the domain G is the strip S with the cuts along the rays r_1 and r_2.

Introduce the notations $E = r_1 \cup r_2$, $\bar{G}$ is the closure of the domain G, ∂G is the boundary of G, $\omega = \omega(z, E, G)$ is the harmonic measure of the set E with respect to the domain G.

Lemma. *For $\omega = \omega(z, E, G)$ the following estimate holds*

$$2/3 < \omega(z, E, G) \leq 1.$$

In the space of the functions $f(x, \alpha)$ we introduce the norm

$$\|f(x_1, x_2, y, \alpha\|_1 = \|f_1(x_1 - x_3 \cos\alpha, x_2 - x_3 \sin\alpha, \alpha)\|_{\mathbf{C}(\mathcal{D})}.$$

Theorem 7. *Suppose the function $u(x, y)$ belongs to the class $C_0^5(\Omega)$ and the inequalities*

$$\|u(\cdot)\|_{C_0^5(\Omega)} < 1, \qquad \|f(\cdot)\|_1 < \varepsilon,$$

hold, where $\varepsilon > 0$ is sufficiently small.

Then the following estimate of conditional stability holds for the solution of the problem of integral geometry (17)

$$\|u(\cdot)\|_{\mathbf{C}} < a_1 |\ln(1/\varepsilon)|^{-1},$$

where a_1 is some constant.

Proof. The stability estimate can be found analogously Theorem 6 by reducing the initial integral geometry problem to the problem of analytical extension of the function $v(\nu_1 + i\nu_2)$ in the strip $\Pi = \{(\nu_1, \nu_2) : \nu_1 \in \mathbf{R}^1$, $|\nu_2| < a\pi$, $0 < a < \infty\}$ from the rays $p_1 = \{(\nu_1, \nu_2) : \nu_1 \geq a, \ \nu_2 = 0\}$ and $p_2 = \{(\nu_1, \nu_2) : \nu_1 \leq -a, \ \nu_2 = 0\}$ into the interval $\{(\nu_1, \nu_2) : -a < \nu_1 < a$, $\nu_2 = 0\}$.

Introduce the domain $\Lambda = \Pi \setminus \{p_1 \cup p_2\}$. The harmonic measure of the set $\{p_1 \cup p_2\}$ with respect to the domain Λ is $\omega = \omega(\nu, p_1 \cup p_2, \Lambda)$.

From the lemma follows the estimate for the harmonic measure ω:

$$2/3 < \omega(\nu, p_1 \cup p_2, \Lambda) \leq 1. \tag{19}$$

The function $v(\nu)$ satisfies the conditions of the theorem on two constants; therefore, taking into account inequality (19), we obtain

$$|v(\nu)| < {\varepsilon_1}^{2/3} {M_1}^{1/3}, \quad \nu \in \Pi, \tag{20}$$

where ε_1 and M_1 are the estimates from above of the modulus of the function $v(\nu, \lambda_2, \lambda_3)$ in the rays p_1 and p_2 and in the lines $\{\nu : \nu_1 \in \mathbf{R}^1, \nu_2 = -a\pi\}$, $\{\nu : \nu_1 \in \mathbf{R}^1, \nu_2 = a\pi\}$ respectively.

In the rays p_1 and p_2 we have

$$|v(\lambda)| < a_2 \varepsilon, \tag{21}$$

where a_2 is a constant.

It is not hard to prove the following estimate for $|v(\cdot)|$ in the boundary of the strip Π:

$$|v(\cdot)| < a_3 e^{a\pi}/(a\pi), \tag{22}$$

where a_3 is a constant.

Estimates (20)–(22) yield the estimate

$$|v| < a_4 a^{-1/3} e^{a\pi/3} \varepsilon^{2/3}, \tag{23}$$

where $a_4 = (a_2^2 a_3/\pi)^{1/3}$.

We obtain now the estimates of $|v(\lambda)|$ inside the ball $\mathcal{G}_R = \{\lambda : |\lambda| < R, 0 < R < \infty\}$ and outside it. Applying the inverse Fourier transformation, we obtain

$$|u(x)| < a_5 \mathcal{H}^{8/3} e^{\pi\mathcal{H}/3} \varepsilon^{2/3} + a_6/\mathcal{H},$$

where a_5, a_6 are constants , $\mathcal{H} = R/\sqrt{2}$.

We obtain now the estimate from above for the sum from the right-hand side of this inequality. To this end, we choose $\mathcal{H}$ so that

$$a_6 \mathcal{H}^{8/3} e^{\pi\mathcal{H}/3} \varepsilon^{2/3} = a_7/\mathcal{H}.$$

Estimating the order of principle part of $\mathcal{H}$, we have

$$|u(x)| < a_1 |\ln(1/\varepsilon)|^{-1}.$$

Theorem 7 is proved. □

Remark 1. All the results obtained above may be easily propagated for the case of generatrices of the cones of more general form

$$\sum_{m=1}^{2} a_m^2 (x_m - \xi_m)^2 (y - \eta)^2,$$

where $a_m \in \mathbf{R}^1$.

Remark 2. We may consider also such generalization of problem of solving equation (17) when the generatruces of cones are smooth closed convex plane curves.

REFERENCES

1. Yu. E. Anikonov, *Some Methods for the Study of Multidimensional Inverse Problems for Differential Equations.* Novosibirsk, Nauka, 1978 (in Russian).

2. Akbar H. Begmatov, One class of problems of integral geometry in the plane. *Dokl. Ros. Akad. Nauk* (1993) **331**, No. 3, 261–262 (in Russian).

3. Akbar H. Begmatov, On certain classes of polysingular integral equations. *Sib. Mat. Zh.* (1994) **35**, No. 3, 515–519.

4. Akbar H. Begmatov, Reductin of the integral geometry problem in the three-dimensinal space to the polysingular integral equation with perturbation. *Dokl. Ros. Akad. Nauk* (1998) **360**, No. 5, 583–585 (in Russian).

5. Akbar H. Begmatov, On some classes of polysingular integral equations. *Sib. Zh. Ind. Mat.* (1999) **2**, No. 2, 8–14 (in Russian).

6. Akbar H. Begmatov, The problem of integral geometry with perturbation in the three-dimensional space. *Sib. Mat. Zh.* (2000) **41**, No. 1, 3–14 (in Russian).

7. Akbar H. Begmatov, One problem of integral geometry with perturbation in the three-dimensional space. *Dokl. Ros. Akad. Nauk* (2000) **371**, No. 2, 155–158.

8. Akbar H. Begmatov, Integral geometry problems of Volterra type. In: *Integral Methods in Science and Engineering.* B. Bertram, C. Constanda and A. Struthers (Eds.). Chapman&Hall/CRC Research Notes

in Mathematics Series, **418**. Chapman&Hall/CRC, Boka Raton, Fl, 2000, 46–50.

9. Akbar H. Begmatov, Problem of integral geometry on paraboloids with perturbation. In: *Proceedings of the Second ISAAC Congress.* Begehr et al. (Eds). Kluwer, Dordrecht, Vol. 1, 97–103.

10. Akbar H. Begmatov, On recovery of surfaces from contours of their shadows. *J. Inv. Ill-Posed Problems* (2002) **10**, No. 3, 213–220.

11. Akram H. Begmatov, Two classes of weakly ill-posed problems of integral geometry in the plane. *Sib. Mat. Zh.* (1995) **36**, No. 2, 243–247 (in Russian).

12. Akram H. Begmatov, Problems of integral geometry for the cone set in n-dimensional space. *Sib. Mat. Zh.* (1996) **37**, No. 3, 500–505 (in Russian).

13. Akram H. Begmatov, Volterra problem of integral geometry in the plane for the curves with singularities. *Sib. Mat. Zh.* (1997) **38**, No. 4, 723–737 (in Russian).

14. Akbar H. Begmatov and Akram H. Begmatov, Inversion of the X-ray transform and the Radon transform with incomplete data, In: *Integral methods in science and engineering.* B. Bertram, C. Constanda and A. Struthers (Eds). Chapman& Hall/CRC Research Notes in Mathematics Series, **418**. Chapman&Hall/CRC, Boka Raton, Fl, 2000, 51–56.

15. A. S. Blagoveshchenskii, On recovery of a function given its integrals over ruled manifolds. *Mat. Zametki* (1986) **39**, No. 6, 841–849.

16. A. L. Bukhgeim, *Volterra Equations and Inverse Problems.* Novosibirsk, Nauka, 1983 (in Russian).

17. A. M. Cormack, Representation of a function by its line integrals, with some radiological applications. *J. Appl. Phys.* (1963) **34**, 2722–2727.

18. A. M. Cormack, Representation of a function by its line integrals, with some radiological applications II. *J. Appl. Phys.* (1964) **35**, 2908–2912.

19. R. Courant, *Partial Differential Equations.* Interscience, New York, 1962.

20. D. V. Finch, Cone beam reconstruction with sources on a curve. *SIAM J. Appl. Math.* (1985) **45**, 665–673.

21. P. Funk, Uber eine geometrische Anwendung der Abelschen Integralgleichung. *Math. Ann.* (1916) **77**, 129–135.

22. I. M. Gel'fand and A. B. Goncharov, Recovery of a compactly-supported function from its integrals over the straight lines intersecting a given set of points in space. *Dokl. Akad. Nauk SSSR* (1986) **290**, 1037–1040.

23. I. M. Gel'fand, S. G. Gindikin, and M. I. Graev, *Some Inverse Problems of Integral Geometry.* Moscow, Dobrosvet, 2000 (in Russian).

24. P. Grangeat, Mathematical framework of cone beam 3D reconstruction via the first derivative of the Radon transform. In: *Mathematical Methods in Tomography.* Herman et al. (Eds). Springer, 1991.

25. S. Helgason, *The Radon Transform.* Birkhauser, Boston, 1980.

26. F. John, Bestimmung einer Funktion aus ihren Integralen über gewisse Mannigfaltigkeiten. *Math. Ann.* (1934) **109**, 488–520.

27. F. John, The ultrahyperbolic differential equation with 4 independent variables. *Duke Math. J.* (1938) **4**, 300–322.

28. F. John, *Plane Waves and Spherical Means Applied to Partial Differential Equations.* Interscience, New York, 1955.

29. A. A. Kirillov, On a problem by I. M. Gel'fand. *Dokl. Akad. Nauk SSSR* (1961) **137**, No. 2, 276–277 (in Russian).

30. M. M. Lavrent'ev, Integral geometry and inverse problems. In: *Ill-Posed Problems of Mathematical Physics and Analysis.* Novosibirsk, Nauka, 1984, 81–86 (in Russian).

31. M. M. Lavrent'ev, Integral geometry problems with perturbation on the plane. *Sib. Mat. Zh.* (1996) **37**, No. 4, 851–857 (in Russian).

32. M. M. Lavrent'ev and V. G. Romanov, Three linearized inverse problems for hyperbolic equations. *Dokl. Akad. Nauk SSSR* (1966) **171**, No. 6, 1279–1281 (in Russian).

33. M. M. Lavrent'ev, V. G. Romanov, and S. P. Shishatskii, *Ill-Posed Problems of Mathematical Physics and Analysis.* Nauka, Moscow, 1980; English transl.: AMS, Providence, Rhode Island, 1986.

34. F. Natterer, *The Mathematics Computerized Tomography.* Teubner, Stuttgart, 1986.

35. F. Natterer, Recent developments in X-ray tomography. In: *Tomography, Impedance Imaging, and Integral Geometry.* Quinto et al. (Eds). AMS, 1994.

36. V. P. Palamodov, Some mathematical aspect of 3D X-ray tomography. In: *Tomography, Impedance Imaging, and Integral Geometry.* Quinto et al. (Eds). AMS, 1994.

37. J. Radon, Über die Bestimmung von Funktionen durch ihre Integralwerte längs gewisser Mannigfaltigkeiten. *Math.-Nat. Kl.* (1917) **69**, 262–277.

38. V. G. Romanov, *Some Inverse Problems for Equations of Hyperbolic Type.* Novosibirsk, Nauka, 1972; English. transl.: *Integral Geometry and Inverse Problems for Hyperbolic equations.* Springer-Verlag, 1974.

39. V. G. Romanov, *Inverse Problems of Mathematical Physics.* Moscow, Nauka, 1984 (in Russian).

40. H. K. Tuy, An inversion formula for cone-beam reconstruction. *SIAM J. Appl. Math.* (1983) **43**, 546–552.

Ill-Posed and Non-Classical Problems of Mathematical Physics and Analysis, Samarkand, 2000, pp. 19–26
M. M. Lavrent'ev and S. I. Kabanikhin (Eds)

Existence of solutions of the first boundary-value problem for the third order equations of mixed type in unbounded domain

T. D. DJURAEV and A. R. HASHIMOV

Abstract — In this paper the solutions of the first boundary-value problem for the third order equation of mixed type are considered. The existence theorem is established in the classes of functions growing in infinity.

Intensive development of theory of generalized functions had allowed to establish the theorems of uniqueness and existence in the classes of increasing functions for linear equations of elliptic and parabolic type for domains with noncompact boundary depending on geometric domain characteristics [1, 2]. For equations of odd order which fill a highly important place in applications the analogous research up to nowadays were absent.

In our paper we consider an unbounded domain $\Omega \subset \mathbb{R}^n_+ = \{x \mid x_1 > 0\}$ and the existence of generalized solution of the problem

$$lAu + Bu = f(x), \tag{1}$$

$$u\big|_{\sigma_0 \cup \sigma_1 \cup \sigma_2} = 0, \qquad l_0 u|_{\sigma_1} = 0, \tag{2}$$

where

$$Au = a^{ij}(x)u_{x_i x_j} + a^i(x)u_{x_i} + a(x)u, \ Bu = b^{ij}(x)u_{x_i x_j} + b^i(x)u_{x_i} + a(x)u,$$
$$lu = l_0 u + \alpha(x)u, \quad l_0 u = \alpha^k(x)u_{x_k}, \quad \Gamma = \partial\Omega,$$
$$\sigma_0 = \{x \in \Gamma \mid \alpha^k(x)\nu_k(x) = 0\}, \quad \sigma_1 = \{x \in \Gamma \mid \alpha^k(x)\nu_k(x) > 0\},$$
$$\sigma_2 = \{x \in \Gamma \mid \alpha^k(x)\nu_k(x) < 0\},$$

$\nu(x) = (\nu_1, \dots \nu_n)$ is unit vector of internal normal to Γ in the point x. Here and further in formulas the summing goes along the repeating indices.

Note that the existence of generalized solution of problem (1), (2) in a bounded domain is researched in [3].

We assume that in a certain neighborhood of each its point the hyperplane Γ is represented in the form $x_j = \chi(x_1, \dots, x_{j-1}, \dots x_{j+1}, \dots, x_n)$ for some j, where χ belongs to the class C^2.

Suggest that all the coefficients in (1), the functions $\alpha(x)$, $\alpha^k(x)$ ($k = 1, 2, \dots, n$), and their derivatives which met further are bounded and measurable in each bounded subdomain of Ω.

We assume further that

$$a^{ij} = a^{ji}, \quad a_0|\xi|^2 \le a^{ij}\xi_i\xi_j \le a_1|\xi|^2, \quad d^{ij} = d^{ji},$$

$$, d_0|\xi|^2 \le d^{ij}\xi_i\xi_j \le d_1|\xi|^2, \quad c_1^{ij} = c_1^{ji}, \quad c_{10}|\xi|^2 \le c_1^{ij}\xi_i\xi_j \le c_{11}|\xi|^2,$$

$$\sum_{k=1}^{n}[\alpha^k(x)]^2 \ne 0, \quad c_1^{ij} - \alpha^i a - c_1^1 < 0,$$

$$q^{ij} = c_1 - c^i_{1x_i}/2 + c^{ij}_{1x_ix_j}/2 + (\alpha^i a)_{x_i}/2 \le -c_0 < 0,$$

for $x \in \Omega \cup \Gamma$, $\xi \in \mathbb{R}^n$; a_0, a_1, d_0, d_1, c_{10}, c_{11}, c_0 are positive constants. Here

$$d^{ij} = c_1^{ij} - (\alpha^i a^{kj})_{x_k} + \alpha^i a^j + (\alpha^k a^{ij})_{x_k}/2,$$

$$c_1^{ij} = b^{ij} + \alpha a^{ij} - \alpha^k_{x_k} a^{ij}, \quad c_1^i = b^i + \alpha a^i - \alpha^k_{x_k} a^i,$$

$$c_1 = b + \alpha a - \alpha^k_{x_k} a.$$

We assume that $\{\Omega_\tau\}$ is the set of bounded subdomains of domain Ω depending on the parameter $\tau \in \Pi = \{\tau \mid 0 \le \tau \le \tau^0\}$, $\tau^0 \le \infty$. We shall assume that $\Omega_\tau \subset \Omega_{\tau'}$ if $\tau < \tau'$. Denote $S_\tau = \partial\Omega_\tau \setminus \partial\Omega_{\tau'}$. We assume that S_τ is connected $(n-1)$-dimensional surface with the same smoothness that $\partial\Omega$ and $\partial S_\tau \subset \partial\Omega$.

Set $\Gamma_\tau = \Gamma \cap \partial\Omega_\tau$, $\sigma_{0,\tau} = \{x \in \Gamma_\tau \mid \alpha^k\nu_k = 0\}$, $\sigma_{1,\tau} = \{x \in \Gamma_\tau \mid \alpha^k\nu_k > 0\}$, $\sigma_{2,\tau} = \{x \in \Gamma_\tau \mid \alpha^k\nu_k < 0\}$. For $h > 0$, we define $\sigma_{1,h,\tau} = \{x \in \sigma_{1,\tau} \mid \rho(x, \partial\sigma_{1\tau}) > h\}$, $\sigma^h_{1,\tau} = \sigma_{1,\tau} \setminus \sigma_{1,h,\tau}$.

Suppose that $E(\Omega_\tau)$ is the set of the functions ϑ from the class $C^2(\Omega_\tau)$ such that $\vartheta = 0$ in Γ_τ and for some $h > 0$ we have $l_0\vartheta = 0$ in $\sigma_{0,\tau} \cup \sigma_{2,\tau} \cup \sigma^h_{1,\tau}$.

Denote by $H(\Omega_\tau)$ the closure of $E(\Omega_\tau)$ in the norm

$$\|\omega\|_{H(\Omega_\tau)} = \Big[\int_{\Omega_\tau} (d^{ij}\omega_{x_i}\omega_{x_j} + \omega^2)\,dx + \int_{\sigma_{1,\tau}} \alpha^k\nu_k a^{ij}\omega_{x_i}\omega_{x_j}\,ds\Big]^{1/2}.$$

Consider the bilinear form

$$a_1(u,\vartheta) = \int_{\Omega_\tau} \big[\alpha^k a^{ij} u_{x_i}\vartheta_{x_j x_k} + (\alpha^k a^{ij})_{x_j} u_{x_i}\vartheta_{x_k} - \alpha^k a_i u_{x_i}\vartheta_{x_k}$$
$$- c_1^{ij} u_{x_i}\vartheta_{x_j} + (c^{ij}_{1x_j} - \alpha^i a - c_1^i)u\vartheta_{x_i} + (c_1 - c^i_{1x_i} + c^{ij}_{1x_i x_j})u\vartheta\big]\,dx. \quad (3)$$

Definition We define *the generalized solution* of the problem (1), (2) in the domain Ω each function $u(x)$ such that for each bounded subdomain Ω_τ of the domain Ω, $u(x) \in H(\Omega_\tau)$ and

$$a_1(u,\vartheta) = \int_{\Omega_\tau} f\vartheta\,dx \quad (4)$$

for any function $\vartheta(x) \in E(\Omega_\tau)$, $\vartheta = 0$ in S_τ.

I. First we suppose that $\alpha^k = \text{const}$, $k = \overline{1,n}$, $\alpha_1 > 0$. In [3] it was shown that the generalized solution satisfies the second condition in (2) on the average.

For simplicity of presentation we assume first that $S_\tau = \Omega \cap \{x \mid x_1 = \tau + \gamma\}$ for each $\tau \in [0,\tau_0]$, $\gamma = \text{const}$.

Introduce the notation

$$Q(u) = d^{ij}u_{x_i}u_{x_j} - q^{ij}u^2, \qquad g = a_1 d_0^{-1/2}(\alpha^1)^2,$$
$$P(\tau) = \sup_{S_\tau} B(x), \quad (5)$$
$$B(x) = \max\{2^{-1}(\alpha^1 a^1 + c_1^{i1} - (\alpha^1 a^{ij})_{x_j}, 0\}. \quad (6)$$

Set

$$0 < \lambda(\tau) \le \inf_{v\in N}\Big\{\int_{S_\tau} Q(v)\,ds\Big[\int_{S_\tau} v^2\,ds\Big]^{-1}\Big\}, \quad (7)$$

where N is a set of functions v continuously differentiable in the neighborhood of S_τ for $x \in \bar\Omega$ and equal to zero in $S_\tau \cap \Gamma$.

We suppose that $\Phi(\tau)$ is positive for $\tau \in \Pi$ and

$$\Phi(\tau) \ge \lambda^{-1/2}(\tau) + P(\tau)\lambda^{-1}(\tau). \quad (8)$$

Depending on the type of the domain the considerations of the work will differ. We shall consider the two classes of domains Ω for which one of the below conditions will hold

A) $\partial\Phi(\tau)/\partial\tau \le \varepsilon$ for all $\tau \in \Pi$, $\varepsilon = \text{const}$, $0 < \varepsilon < 1$;

B) $\partial\Phi(\tau)/\partial\tau \ge \varepsilon$ for all $\tau \in \Pi$

and satisfying the condition $\tau(0) = 0$. Here the function Φ is such that the right-hand sides of the below formulas (9), (10) will be absolutely continuous.

For problem (1), (2) the following theorems hold [4].

Theorem 1 [the analog of the Sen-Venan principle]. *Suppose that $u(x)$ is the generalized solution of problem* (1), (2) *in the domain Ω of the class), where*

$$(\alpha^1 a^i)_{x_i} - (\alpha^1 a^{ij})_{x_i x_j} + 3c^{ij}_{1x_i x_j} - 2\alpha^i a - 2c^1_1 \geq 0 \text{ and } f(x) = 0 \text{ in } \Omega_{\tau^0}. \quad (9)$$

Then for each R_0 and R such that $0 \leq R_0 \leq R$ the following estimate holds

$$\int_{\Omega_{\tau(R_0)}} Q(u)\,dx \leq \frac{\tau(R)+1}{\tau(R_0)+1} \exp\{-(R-R_0)\} \int_{\Omega_{\tau(R)}} Q(u)\,dx. \quad (10)$$

Theorem 2 [the analog of the Sen-Venan principle]. *Suppose that $u(x)$ is the generalized solution of problem* (1), (2) *in the domain Ω of the class B), and condition* (9) *holds. Then for each R_0 and R such that $0 \leq R_0 \leq R$ the following estimate holds*

$$\int_{\Omega_{\tau(R_0)}} Q(u)\,dx \leq \exp\Big\{\int_{\tau(R_0)}^{\tau(R)} \frac{\varepsilon s\,ds}{\Phi(s)}\Big\} \int_{\Omega_{\tau(R)}} Q(u)\,dx. \quad (11)$$

Now, we establish the existence of solution of problem (1), (2).

Lemma 1. *We suppose that there exists an infinite sequence of bounded subdomains $\Omega_{\tau(N)}$ of the domain Ω from the class) such that $\Omega_{\tau(N)} \subset \Omega_{\tau(N+1)}$ for $N = 1, 2, \ldots, \Omega = \bigcup_{i=1}^n \Omega_{\tau(i)}$. Suppose that for each fixed N for each function ω which is a generalized solution of equation (1) in $\bar{\Omega}_{\tau(N+1)} \cap \Gamma$ the following estimate holds*

$$\int_{\Omega_{\tau(N)}} Q(\omega)\,dx \leq \frac{\tau(N+1)+1}{\tau(N)+1} \exp^{-1} \int_{\Omega_{\tau(N+1)}} Q(\omega)\,dx. \quad (12)$$

Suppose also the the function $f(x)$ is defined in Ω and satisfies the estimate

$$\int_{\Omega_{\tau(N)}} f^2\,dx \leq M_1 \Lambda(\Omega_{\tau(N)}) \exp[(1-\delta)N], \qquad n = 1, 2, \ldots, \quad (13)$$

where $\delta = \text{const}$, $0 < \delta < 1$, the constant M_1 is independent of N,

$$\Lambda(\Omega_{\tau(N)}) = \inf_{\vartheta \in H(\Omega_\tau)} \Big\{\int_{\Omega_{\tau(N)}} Q_(\vartheta)\,dx \Big(\int_{\Omega_{\tau(N)}} \vartheta^2\,dx\Big)^{-1}\Big\}. \quad (14)$$

Then there exists a unique generalized solution $u(x)$ of problem (1), (2) in Ω which satisfies the estimate

$$\int_{\Omega_{\tau(N)}} Q(\omega)\,dx \le M_2 \exp\,[(1-\delta)N], \qquad N = 1, 2, \ldots, \tag{15}$$

where M_2 is the constant independent of N.

Proof. Note that $\Lambda(\Omega_{\tau(N)}) > 0$. Denote by $u_l^m(x)$ the sequence of functions from $E(\Omega_l)$ such that u_l^m converges to u_l for $m \to \infty$ in the norm $H(\Omega_l)$; u_l is the generalized solution of problem (1), (2) in the domain Ω_l [3]. We fix an arbitrary subdomain $\Omega_{\tau(N)}$ from the sequence $\Omega_{\tau(1)} \subset \Omega_{\tau(2)} \subset \ldots$ and consider the sequence of subdomains $\Omega_{\tau(N+k)}$, $k \to \infty$.

Taking into account (4), for $\Omega_\tau = \Omega_{\tau(N+k)}$, $u = u_{N+k}$, $\vartheta = u^m_{N+k}$, integrating by parts and passing to the limit for $m \to \infty$ we obtain

$$\int_{\Omega_{\tau(N+k)}} Q(u_{N+k})\,dx \le -\int_{\Omega_{\tau(N+k)}} f \cdot u_{N+k}\,dx.$$

Hence, taking into account the Cauchy–Bunyakovskii inequality, we have

$$\int_{\Omega_{\tau(N+k)}} Q(u_{N+k})\,dx \le \Big(\int_{\Omega_{\tau(N+k)}} f^2\,dx\Big)^{1/2} \Big(\int_{\Omega_{\tau(N+k)}} u^2_{N+k}\,dx\Big)^{1/2},$$

$$\int_{\Omega_{\tau(N+k)}} Q(u_{N+k})\,dx \le \Lambda^{-1}(\Omega_{\tau(N+k)}) \int_{\Omega_{\tau(N+k)}} f^2\,dx.$$

Conditions (13) and this inequality yield

$$\int_{\Omega_{\tau(N+k)}} Q(u_{N+k})\,dx \le M_1 \exp\,\{(1-\delta)(N+k)\}. \tag{16}$$

Set

$$\|\vartheta\|_{\Omega_l} = \Big(\int_{\Omega_l} Q(\vartheta)\,dx\Big)^{1/2}.$$

It is easy to see that the function $\omega = u_{N+k+1} - u_{N+k}$ in the domain Ω_{N+k} is the generalized solution of equation (1) for $f(x) = 0$ satisfying conditions (2). Therefore, applying subsequently estimate (12) to the domains $\Omega_{\tau(N)}, \Omega_{\tau(N+1)}, \ldots, \Omega_{\tau(N+k)}$ and taking into account (16) we find

$$\|u_{N+k+1} - u_{N+k}\|_{\Omega_{\tau(N)}} \le M_3 \Big(\frac{\tau(N+1)+1}{\tau(N)+1}\Big)^{k/2} e^{-\delta k/2}, \tag{17}$$

where the constant M_3 is independent of N and k.

Inequality (17) for each natural $p > 0$ and $q > 0$ yields

$$\|u_{N+p+q} - u_{N+p}\|_{\Omega_{\tau(N)}} \leq M_4\Big(\frac{\tau(N+1)+1}{\tau(N)+1}\Big)^{p/2} e^{-\delta p/2}, \tag{18}$$

where the constant M_4 is independent of N, p, and q. Therefore $\|u_{N+k+1} - u_{N+k}\|_{\Omega_{\tau(N)}} \to 0$ for each $q > 0$ and $p \to \infty$.

Since $H(\Omega_{\tau(N)})$ is a Hilbert space, then

$$\|u_{N+p} - u_{N+p+q}\|^2_{H(\Omega_{\tau(N)})} \leq M_5\|u_{N+p} - u_{N+p+q}\|^2_{\Omega_{\tau(N)}},$$

where the constant M_5 is independent of p. Hence it follows that the sequence $\{u_p\}$ converges in the norm $H(\Omega_{\tau(N)})$ for $p \to \infty$ to the function $u(x) \in H(\Omega_{\tau(N)})$. As $N > 0$ is taken arbitrarily, then $u(x)$ is defined in Ω and $u(x) \in H(\Omega')$ for each bounded subdomain Ω' of the domain Ω.

Taking into account the imbedding theorems, the trace of the function $u(x)$ in $\partial\Omega' \cap \Gamma$ exists and the first condition from (2) holds in this set. As we have noted the second condition from(2) holds following [3].

It is easy to see that u_{N+p_m} satisfies integral identity (4) when $\Omega' = \Omega_{\tau(N)}$ and $\vartheta(x) \in E(\Omega_{\tau(N)})$. Passing to the limit for $p_m \to \infty$ we see that $u(x)$ satisfies integral identity (4) for $\Omega_\tau = \Omega_{\tau(N)}$ and $\vartheta \in E(\Omega_N)$. Since $N > 0$ is arbitrary, then, for each subdomain $\Omega_N \subset \Omega$ and the function $\vartheta \in E(\Omega_N)$, integral identity (4) holds. Therefore, $u(x)$ is the generalized solution of problem (1), (2) in the domain Ω.

Setting in relation (18) $p = 1$, tending q to ∞ , and taking into account estimate (16) we find

$$\begin{aligned}\|u\|_{\Omega_{\tau(N)}} &\leq k \exp\{(1-\delta)N\} + \|u_{N+1}\|_{\Omega_N} \\ &\leq k \exp\{(1-\delta)N/2\} + k_1 \exp\{(1-\delta)(N+1)/2\}.\end{aligned}$$

Hence relation (15) follows.

We show now that the solution $u(x)$ of problem (1), (2) satisfying conditions (15) is unique. As it follows from inequality (12), for each two generalized solutions u and ϑ of problem (1), (2) in the domain Ω for which estimate (15) holds the following estimate holds for each $l > 0$

$$\begin{aligned}&\int_{\Omega_{\tau(l)}} Q(u-\vartheta)\,dx \leq 2^j M_3 \exp(-j) \int_{\Omega_{\tau(l+j)}} Q(u-\vartheta)\,dx \\ &\leq 2^j M_3 \exp\{-j\}\Big[\int_{\Omega_{\tau(l+j)}} Q(u)\,dx + \int_{\Omega_{l+j}} Q(\vartheta)\,dx\Big] \leq 2^j M_6 \exp(-\delta j),\end{aligned}$$

where the constant M_6 is independent of j. Passing to the limit when $j \to \infty$ in this estimate, we obtain $\|u - \vartheta\|_{\Omega_{\tau(l)}} = 0$ for all $l > 0$. Since $\Lambda(\Omega_{\tau(l)}) > 0$ then $u - \vartheta \equiv 0$ in $\Omega_{\tau(l)}$; therefore, $u \equiv \vartheta$ in Ω. □

As it follows from this lemma, the existence of the generalized solution of problem (1), (2) in Ω is based on the assumption that relation (12) holds.

We show now how to construct the sequence of domains $\{\Omega_{\tau(N)}\}$ so that relations (12) holds. Suppose $\{\Omega_\tau\}$ is a set of bounded subdomains of the domain Ω and let all the conditions of Theorem 1 hold. Then, for each R_0 and R such that $0 \le R_0 \le R$ the following estimate holds

$$\int_{\Omega_{\tau(R_0)}} Q(u)\,dx \le \frac{\tau(R)+1}{\tau(R_0)+1}\,e^{-(R-R_0)} \int_{\Omega_{\tau(R)}} Q(u)\,dx. \tag{19}$$

The lemma and estimate (19) yield the following theorem of existence and uniqueness for generalized solution (1), (2).

Theorem 3. *Suppose for each domain $\Omega_{\tau(R)}$ we have $\Lambda(\Omega_{\tau(k)}) > 0$; the function $f(x)$ is defined in Ω and satisfies the relations*

$$\Lambda^{-1}(\Omega_{\tau(k)} \int_{\Omega_{\tau(k)}} f^2\,dx \le M_2 \exp(1-\delta)k, \qquad k = 1, 2, \dots,$$

where $\delta = \text{const}$, $0 < \delta < 1$, the constant M_2 is independent of k, $\tau(k)$ is the function defined in (19). Then there exists a unique generalized solution $u(x)$ of problem (1), (2) and the inequalities hold

$$\int_{\Omega_{\tau(k)}} Q(u)\,dx \le M_7 \exp\{(1-\delta)k\}, \qquad k = 1, 2, \dots.$$

Proof. Set $\Omega_{\tau(N)} = \Omega_{\tau(j)}$ in Lemma 1. Inequality (12) for the functions $u_i - u_{i-1}$ follows from Theorem 1. Thus, Theorem 3 follows from Lemma 1. □

For the domains of the class B) the following theorem holds.

Theorem 4. *Suppose for each domain $\Omega_{\tau(R)}$ we have $\Lambda(\Omega_{\tau(k)}) > 0$ and $f(x)$ defined in Ω satisfies the relations*

$$\Lambda^{-1}(\Omega_{\tau(k)}) \int_{\Omega_{\tau(k)}} f^2\,dx \le M_8 \exp\left\{(1-\delta)\int_0^{\tau(k)} \frac{\varepsilon s\,ds}{\Phi(s)}\right\}, \quad k = 1, 2, \dots,$$

where $\delta = \text{const}$, $0 < \delta < 1$, $\varepsilon = \text{const}$, $0 < \varepsilon < 1$; the constant M_8 is independent of k; $\tau(k)$ is the function defined in (11). Then there exists *a unique generalized solution $u(x)$ of problem* (1), (2) *which satisfies the inequalities*

$$\int_{\Omega_{\tau(k)}} Q(u)\,dx \le M_9 \exp\left\{(1-\delta)\int_0^{\tau(k)} \frac{\varepsilon s\,ds}{\Phi(s)}\right\}, \quad k = 1, 2, \ldots .$$

Theorem 4 is proved analogously Theorem 3.

II. Suppose $\alpha^k(x)\nu_k(x) < 0$ in Γ. Then problem (1), (2) becomes the Dirichlet problem; the operator l_0, in this case, has the variable coefficients and the solution $u(x)$ of problem (1), (2) belongs to the space $\tilde{W}_2^2(\Omega_\tau)$ [3]. In this case we can obtain the analog of the Sen-Venan principle and the existence theorems. In the classes of functions growing in infinity similarly as in point **I**. The proof of these statement is just the same as in point **I**.

Remark. The analogous results hold when $S_\tau = \Omega \cap \{|x| = \tau + \gamma\}$.

REFERENCES

1. O. A. Oleinik and G. Ȧ. Iosif'ean. *Uspekhi Mat. Nauk* (1976) **31**, No. 113, 142–166 (in Russian).

2. O. A. Oleinik and G. Ȧ. Iosif'ean. *Mat. Sbornik* (1980) **4**, 588–610 (in Russian).

3. A. I. Kojanov, *Boundary-Value Problems of the Odd Order for Mathematics Physics.* Nauka, Novosibirsk, 1990 (in Russian).

4. A. R. Hashimov. *Uzb. Mat. Zh.* (2001) **5–6**, 64–71.

Ill-Posed and Non-Classical Problems of Mathematical Physics and Analysis, Samarkand, 2000, pp. 27–55
M. M. Lavrent'ev and S. I. Kabanikhin (Eds)

On the monotone error rule for choosing the regularization parameter in ill-posed problems

U. HÄMARIK* and U. TAUTENHAHN†

Abstract — We consider linear ill-posed problems $Ax = y$ in Hilbert spaces with minimum-norm solution $x^\dagger$ and suppose that instead of y noisy data y^δ are given satisfying $\|y - y^\delta\| \le \delta$ with known noise level δ. For the stable numerical solution regularization methods are considered including continuous regularization methods such as ordinary Tikhonov regularization $x_r^\delta = (A^*A + r^{-1}I)^{-1}A^*y^\delta$ and iterative regularization methods. For the proper choice of the regularization parameter r (which is the stopping index in iterative methods) we study the *monotone error rule* (ME rule): Choose $r = r_{ME}$ as the largest r-value for which it can be guaranteed that the error $\|x_r^\delta - x^\dagger\|$ is monotonically decreasing for $r \in (0, r_{ME}]$. We compare this rule with other *a posteriori* rules and give conditions for which convergence and order optimal convergence rate results can be guaranteed. For the computation of r_{ME} in Tikhonov methods some nonlinear equation has to be solved. Newton's iteration for this equation appears to be globally and monotonically convergent. Numerical experiments are provided that verify some of the theoretical results.

1. INTRODUCTION

In this paper we consider linear ill-posed problems

$$Ax = y \tag{1.1}$$

*University of Tartu, Department of Mathematics, Liivi 2, 50409 Tartu, Estonia. E-mail: Uno.Hamarik@math.ut.ee

†University of Applied Sciences Zittau/Görlitz, Department of Mathematics, P.O. Box 1455, 02755 Zittau, Germany. E-mail: u.tautenhahn@hs-zigr.de

The work was supported by Deutsche Forschungsgemeinschaft (grant 436 EST) and by the Estonian Science Foundation (grant 4345)

where $A \in \mathcal{L}(X,Y)$ is a bounded operator with non-closed range $R(A)$ and X, Y are infinite dimensional real Hilbert spaces with inner products $(\cdot,\cdot)$ and norms $\|\cdot\|$, respectively. We are interested in the minimum-norm solution $x^\dagger$ of problem (1.1) and assume that instead of *exact* data y there are given *noisy* data $y^\delta \in Y$ with $\|y - y^\delta\| \le \delta$ and known noise level δ.

Ill-posed problems (1.1) arise in a wide variety of problems in applied sciences. For their stable numerical solution regularization methods are necessary, see [4, 19, 26, 39]. Regularization methods can be divided into continuous regularization methods that include ordinary Tikhonov regularization $x_r^\delta := R_r y^\delta = (A^*A + r^{-1}I)^{-1}A^*y^\delta$, and iterative regularization methods where the stopping index plays the role of the regularization parameter. The element $x_r^\delta = R_r y^\delta$ is called regularized approximation for the minimum-norm solution $x^\dagger$ of problem (1.1) provided

1. for any $r > 0$, $R_r : Y \to X$ is continuous and
2. for arbitrary $y \in Y$ with $Qy \in R(A)$, $\lim_{r\to\infty} \|R_r y - x^\dagger\| = 0$, where Q is the projection operator onto $\overline{R(A)}$

(see [36]). Traditional regularization methods possess the property that in the case of exact data the error $\|x_r^0 - x^\dagger\|$ as a function of r is monotonically decreasing for $r \to \infty$. This property is no longer true for the error $\|x_r^\delta - x^\dagger\|$. In the case of noisy data the situation is as follows:

(i) If r becomes too large, then the error $\|x_r^\delta - x^\dagger\|$ increases due to the fact of ill-posedness of the operator equation (1.1).

(ii) If r becomes too small, then the error $\|x_r^\delta - x^\dagger\|$ increases due to the fact that the error $\|x_r^0 - x^\dagger\|$ is monotonically increasing for decreasing r-values.

The monotone decrease of the error $\|x_r^\delta - x^\dagger\|$ for growing r-values can only be guaranteed for small r. Typically $\|x_r^\delta - x^\dagger\|$ diverges for $r \to \infty$. Therefore a rule for the proper choice of the regularization parameter r is necessary.

In the monotone error rule for choosing a proper regularization parameter the idea consists in *searching for a largest computable regularization parameter $r = r_{ME}$ for which we can guarantee that the error $\|x_r^\delta - x^\dagger\|$ is monotonically decreasing for $r \in (0, r_{ME}]$.* For continuous regularization methods this means that

$$\frac{\mathrm{d}}{\mathrm{d}r}\|x_r^\delta - x^\dagger\|^2 \le 0 \quad \text{for all} \quad r \in (0, r_{ME}], \tag{1.2}$$

for iteration methods this means that

$$\|x_r^\delta - x^\dagger\| \le \|x_{r-1}^\delta - x^\dagger\| \quad \text{for all} \quad r = 1, 2, \ldots, r_{ME}\,. \tag{1.3}$$

Similar rules for the choice of the regularization parameter which are based on monotonicity properties of the error were proposed and studied for some iterative methods in [1, 2, 13, 15, 16], for the method of ordinary Tikhonov regularization in [34] and for the method of iterated Tikhonov regularization and some other continuous regularization methods in [14, 15, 35].

In this paper we study the ME rule for continuous regularization methods including Tikhonov methods and asymptotical regularization, and for iterative regularization methods including gradient type methods (Landweber's method, steepest descent method, minimal error method) and implicit iteration methods. We compare the ME rule with other *a posteriori* rules including Morozov's discrepancy principle and study questions concerning convergence $x_{r_{ME}}^\delta \to x^\dagger$ for $\delta \to 0$ and concerning order optimal error bounds under certain source conditions. For the computation of r_{ME} in Tikhonov methods some nonlinear equation $\tilde{d}_{ME}(r) = \delta$ has to be solved. The function $\tilde{d}_{ME}$ appears to be strictly monotonically decreasing and strictly convex. These properties guarantee global and monotone convergence for Newton's iteration. In the final section numerical examples are provided which verify some of the theoretical results.

2. CONTINUOUS REGULARIZATION METHODS

2.1. Continuous regularization methods and the ME rule

In continuous regularization methods which include the method of ordinary Tikhonov regularization $x_r^\delta = (A^*A + r^{-1}I)^{-1}A^*y^\delta$ we use in this section for the regularization parameter the traditional notation $\alpha = 1/r$ instead of r. We consider continuous regularization methods of the general form

$$x_\alpha^\delta = g_\alpha(A^*A)A^*y^\delta\,. \tag{2.1}$$

Here $g_\alpha(\lambda) : [0, a] \to \mathbb{R}$ with $a = \|A\|^2$ is a family of piecewise continuous functions depending on a positive regularization parameter $\alpha > 0$ and the operator function g_α is defined according to

$$g_\alpha(A^*A) = \int_0^a g_\alpha(\lambda)\,\mathrm{d}E_\lambda$$

where $A^*A = \int_0^a \lambda\,\mathrm{d}E_\lambda$ is the spectral decomposition of the operator A^*A. For the functions $g_\alpha(\lambda)$ we assume as in [4, 26, 38, 39] that there exist constants γ, γ_p and p_0 such that for $\alpha > 0$

$$\sup_{0 \le \lambda \le a} |g_\alpha(\lambda)| \le \gamma \alpha^{-1} \tag{2.2}$$

and

$$\sup_{0 \le \lambda \le a} \lambda^p |1 - \lambda g_\alpha(\lambda)| \le \gamma_p \alpha^p \quad \text{for} \quad 0 \le p \le p_0 \,. \tag{2.3}$$

The largest constant p_0 in assumption (2.3) is called *qualification* of the regularization method (2.1) (see [39]). Three well known *a posteriori* rules for choosing the regularization parameter α in continuous regularization methods (2.1) are:

1. *Morozov's discrepancy principle* [4, 27, 38, 39]. In this principle (*D principle*) the parameter $\alpha = \alpha_D$ is chosen as the solution of the equation
$$d_D(\alpha) := \|y^\delta - Ax_\alpha^\delta\| = C\delta \quad \text{with} \quad C \ge 1 \,.$$

2. *Rule of Raus* [30]. In this rule, which we call *R rule*, the regularization parameter $\alpha = \alpha_R$ is chosen as the solution of the equation
$$d_R(\alpha) := \|(I - g_\alpha(AA^*)AA^*)^{1/(2p_0)}(y^\delta - Ax_\alpha^\delta)\| = C\delta \quad \text{with} \quad C \ge 1 \,.$$
Here p_0 is the (largest) constant from assumption (2.3).

3. *Rule of Engl and Gfrerer* [3]. In this rule, which we call *EG rule*, the regularization parameter $\alpha = \alpha_{EG}$ is chosen as the solution of the equation
$$d_{EG}(\alpha) := \gamma^{-1/2} \alpha \Big(Ax_\alpha^\delta - y^\delta, \frac{\mathrm{d}}{\mathrm{d}\alpha} g_\alpha(AA^*) y^\delta \Big)^{1/2} = C\delta \quad \text{with} \quad C \ge 1 \,.$$
Here γ is the (smallest) constant from assumption (2.2).

For ordinary and iterated Tikhonov methods the R– and EG rules coincide. The resulting rule was also proposed in [8] and we call this rule *Raus—Gfrerer rule* (*RG rule*).

Now let us turn over to the ME rule. The general idea of this rule (see Chapter 1) consists in searching for a largest computable $r = r_{ME}$ for which we can guarantee that the error $\|x_r^\delta - x^\dagger\|$ is monotonically decreasing for $r \in (0, r_{ME}]$. The reformulation of this idea in terms of $\alpha = 1/r$ means: *Search for the smallest computable regularization parameter* $\alpha = \alpha_{ME}$ *for which we can guarantee that*

$$\frac{\mathrm{d}}{\mathrm{d}\alpha} \|x_\alpha^\delta - x^\dagger\|^2 \ge 0 \quad \text{for all} \quad \alpha \in [\alpha_{ME}, \infty) \,.$$

In order to guarantee this property we use the identity $g_\alpha(A^*A)A^* = A^*g_\alpha(AA^*)$ and estimate the derivative of the squared error $\|x_\alpha^\delta - x^\dagger\|^2$ with respect to α as follows:

$$\begin{aligned}\frac{1}{2}\frac{\mathrm{d}}{\mathrm{d}\alpha}\|x_\alpha^\delta - x^\dagger\|^2 &= \Big(x_\alpha^\delta - x^\dagger, A^*\frac{\mathrm{d}}{\mathrm{d}\alpha}g_\alpha(AA^*)y^\delta\Big)\\ &= \Big(Ax_\alpha^\delta - y^\delta + (y^\delta - y), \frac{\mathrm{d}}{\mathrm{d}\alpha}g_\alpha(AA^*)y^\delta\Big)\\ &\geq \Big\|\frac{\mathrm{d}}{\mathrm{d}\alpha}g_\alpha(AA^*)y^\delta\Big\|\left\{\frac{\Big(Ax_\alpha^\delta - y^\delta, \frac{\mathrm{d}}{\mathrm{d}\alpha}g_\alpha(AA^*)y^\delta\Big)}{\Big\|\frac{\mathrm{d}}{\mathrm{d}\alpha}g_\alpha(AA^*)y^\delta\Big\|} - \delta\right\}.\end{aligned}$$

This estimate leads us to following ME rule for continuous regularization method (2.1):

ME rule. For regularization methods with monotonically increasing functions $d_{ME}(\alpha)$ in (2.4), choose $\alpha = \alpha_{ME}$ as the solution of the equation

$$d_{ME}(\alpha) := \Big\|\frac{\mathrm{d}}{\mathrm{d}\alpha}g_\alpha(AA^*)y^\delta\Big\|^{-1}\Big(Ax_\alpha^\delta - y^\delta, \frac{\mathrm{d}}{\mathrm{d}\alpha}g_\alpha(AA^*)y^\delta\Big) = \delta. \quad (2.4)$$

2.2. The ME rule for ordinary and iterated Tikhonov regularization

In these methods we start with $x_{\alpha,0}^\delta = 0$ and compute the regularized solution $x_\alpha^\delta := x_{\alpha,m}^\delta$ recursively by solving the m operator equations

$$(A^*A + \alpha I)x_{\alpha,k}^\delta = A^*y^\delta + \alpha x_{\alpha,k-1}^\delta, \quad k = 1, 2, \ldots, m. \quad (2.5)$$

For $m = 1$ this method is the *method of ordinary Tikhonov regularization.* In these methods we have $g_\alpha(\lambda) = [1 - (1 + \lambda/\alpha)^{-m}]/\lambda$ with some fixed positive integer $m \geq 1$. Let $r_{\alpha,m}$ denote the discrepancy of the regularized solution $x_\alpha^\delta = x_{\alpha,m}^\delta$, i.e.

$$r_{\alpha,m} = y^\delta - Ax_{\alpha,m}^\delta.$$

Using the identities

$$1 - \lambda g_\alpha(\lambda) = \Big(\frac{\alpha}{\lambda+\alpha}\Big)^m \quad \text{and} \quad \frac{\mathrm{d}}{\mathrm{d}\alpha}g_\alpha(\lambda) = -\frac{m}{\alpha^2}\Big(\frac{\alpha}{\lambda+\alpha}\Big)^{m+1}$$

we obtain

$$r_{\alpha,m} = [I - AA^*g_\alpha(AA^*)]y^\delta = [\alpha(AA^* + \alpha I)^{-1}]^m y^\delta$$

and

$$\frac{\mathrm{d}}{\mathrm{d}\alpha} g_\alpha(AA^*)y^\delta = -\frac{m}{\alpha^2} r_{\alpha,m+1} \,.$$

From these representations we conclude that the functions $d_{RG}(\alpha)$ and $d_{ME}(\alpha)$ for the RG– and ME rules have the form

$$d_{RG}(\alpha) = (r_{\alpha,m}, r_{\alpha,m+1})^{1/2} \quad \text{and} \quad d_{ME}(\alpha) = \frac{(r_{\alpha,m}, r_{\alpha,m+1})}{\|r_{\alpha,m+1}\|} \,. \tag{2.6}$$

The function $d_{ME}(\alpha)$ given in (2.6) possesses following properties, see [34] for $m = 1$ and [14, 35] for $m \geq 1$:

Theorem 2.1. *Let P denote the orthoprojection of Y onto $N(A^*) = \overline{R(A)}^\perp$ and let $A^* y^\delta \neq 0$. Then:*

(i) *$d_{ME}(\alpha)$ is strictly monotonically increasing and obeys*

$$d_{ME}(0) = \|Py^\delta\| \quad \text{and} \quad \lim_{\alpha\to\infty} d_{ME}(\alpha) = \|y^\delta\| \,.$$

The equation $d_{ME}(\alpha) = \delta$ has a unique solution α_{ME} provided $\|Py^\delta\| < \delta < \|y^\delta\|$.

(ii) *For all $\alpha \in (\alpha_{ME}, \infty)$ there holds $\frac{\mathrm{d}}{\mathrm{d}\alpha} \|x^\delta_{\alpha,m} - x^\dagger\|^2 > 0$.*

(iii) *There holds*

$$d_{RG}(\alpha) < d_{ME}(\alpha) < d_D(\alpha) \,.$$

If $C = 1$ in the D principle and in the RG rule, then $\alpha_D < \alpha_{ME} < \alpha_{RG}$.

From parts (ii) and (iii) of the theorem there follows

$$\|x^\delta_{\alpha_{ME}} - x^\dagger\| < \|x^\delta_{\alpha_{RG}} - x^\dagger\| \,. \tag{2.7}$$

Hence, the ME rule provides always a smaller error than the RG rule. Exploiting the monotonicity property (ii) we obtain for the parameter choice $\alpha = \alpha_{ME}$ order optimal error bounds (see [35]):

Theorem 2.2. *Assume $x^\dagger = (A^*A)^{p/2} w$ with $\|w\| \leq E$. Then for $p \in (0, 2m]$*

$$\|x^\delta_{\alpha_{ME}} - x^\dagger\| \leq \{2^{p/(p+1)} + 2^{-1/(p+1)} \gamma_* (\gamma_{p/2})^{1/p}\} \, E^{1/(p+1)} \, \delta^{p/(p+1)} \tag{2.8}$$

with $\gamma_* = \sqrt{m}$ and $\gamma_{p/2} = [p/(2m)]^{p/2} [1 - p/(2m)]^{m-p/2} \leq 1$.

2.3. The ME rule for asymptotical regularization

In this regularization method the regularized solution is given by $x_\alpha^\delta = x^\delta(r)$ where $x^\delta(r)$ is the solution of the initial value problem

$$\frac{\mathrm{d}}{\mathrm{d}t} x^\delta(t) + A^*Ax^\delta(t) = A^*y^\delta \quad \text{for} \quad 0 < t \leq r\,, \quad x^\delta(0) = 0$$

with $r = 1/\alpha$. In this method we have $g_\alpha(\lambda) = (1 - e^{-\lambda/\alpha})/\lambda$. Using the identities

$$\frac{\mathrm{d}}{\mathrm{d}\alpha} g_\alpha(\lambda) = -\frac{1}{\alpha^2} e^{-\lambda/\alpha} = \frac{1}{\alpha^2}(1 - \lambda g_\alpha(\lambda))$$

we obtain that for the method of asymptotical regularization there holds

$$d_R(\alpha) = d_{EG}(\alpha) = d_{ME}(\alpha) = d_D(\alpha)\,.$$

From the identity $d_D(\alpha) = d_{ME}(\alpha)$ and the monotonicity property of our ME rule we conclude that for the method of asymptotical regularization the best constant C in the D principle is $C = 1$:

Theorem 2.3. *Let $A^*y^\delta \neq 0$, let x_α^δ be the regularized solution obtained by the method of asymptotical regularization and let α_D be the regularization parameter of the D principle with $C = 1$. Then*

$$\|x_{\alpha_D}^\delta - x^\dagger\| < \|x_\alpha^\delta - x^\dagger\| \quad \textit{for all} \quad \alpha > \alpha_D\,.$$

From the identity $d_R(\alpha) = d_{EG}(\alpha) = d_{ME}(\alpha) = d_D(\alpha)$ we conclude that all results known for the D principle (see, e.g., [4, 38, 39]) are also true for the R rule, the EG rule and the ME rule, respectively. Exploiting the monotonicity property of our ME rule we obtain order optimal error bounds for $\|x_\alpha^\delta - x^\dagger\|$ with α chosen by the ME rule, or equivalently, the R rule, the EG rule or the D principle, respectively.

Theorem 2.4. *Assume $x^\dagger = (A^*A)^{p/2}w$ with $\|w\| \leq E$. Let x_α^δ the regularized solution obtained by the method of asymptotical regularization and let α be chosen by the ME rule. Then for all $p \in (0, \infty)$ the order optimal error estimate (2.8) holds true with $\gamma_* \approx 0.6382$ and $\gamma_{p/2} = (p/(2e))^p$.*

Proof. Let us introduce the two operators

$$K_\alpha = I - A^*Ag_\alpha(A^*A)\,, \quad \tilde{K}_\alpha = I - AA^*g_\alpha(AA^*)\,.$$

Then we obtain from (2.1)

$$x_\alpha^\delta - x^\dagger = -K_\alpha x^\dagger + g_\alpha(A^*A)A^*(y^\delta - Ax^\dagger)\,, \tag{2.9}$$

$$Ax_\alpha^\delta - y^\delta = -\tilde{K}_\alpha A x^\dagger + \tilde{K}_\alpha (Ax^\dagger - y^\delta)\,. \tag{2.10}$$

The method of asymptotical regularization is characterized by $g_\alpha(\lambda) = (1 - e^{-\lambda/\alpha})/\lambda$. For this function there holds for arbitrary $\alpha > 0$ the estimate

$$\sup_{0 \le \lambda \le \alpha} \sqrt{\lambda}\, |g_\alpha(\lambda)| \le \gamma_* / \sqrt{\alpha} \tag{2.11}$$

with $\gamma_* \approx 0.6382$ (see [38, 39]). From (2.9) and (2.11) we obtain

$$\|x_\alpha^\delta - x^\dagger\| \le \|K_\alpha x^\dagger\| + \gamma_* \delta / \sqrt{\alpha}\,. \tag{2.12}$$

Now we exploit the assumption $x^\dagger = (A^*A)^{p/2} w$ with $\|w\| \le E$ and obtain from (2.3), which holds with $p_0 = \infty$ and $\gamma_p = (p/e)^{2p}$, the estimate

$$\|K_\alpha x^\dagger\| \le \gamma_{p/2} \alpha^{p/2} E\,. \tag{2.13}$$

For arbitrary $D = D^* \ge 0$ and $0 \le s \le t$ there holds the moment inequality

$$\|D^s v\| \le \|D^t v\|^{s/t} \|v\|^{1-s/t} \tag{2.14}$$

(see, e.g., [22]). We apply this inequality with $D = (A^*A)^{1/2}$, $v = K_\alpha w$, $s = p$, $t = p+1$ and obtain due to the relations $D^{p+1} K_\alpha w = \tilde{K}_\alpha A x^\dagger$ and $\|K_\alpha\| \le 1$ that

$$\begin{aligned} \|K_\alpha x^\dagger\| &= \|D^p K_\alpha w\| \\ &\le \|D^{p+1} K_\alpha w\|^{p/(p+1)} \|K_\alpha w\|^{1/(p+1)} \\ &\le \|\tilde{K}_\alpha A x^\dagger\|^{p/(p+1)} E^{1/(p+1)}\,. \end{aligned} \tag{2.15}$$

For $\alpha = \alpha_{ME} = \alpha_D$ we obtain due to the relations $\|Ax_{\alpha_{ME}}^\delta - y^\delta\| = \delta$, $\|Ax^\dagger - y^\delta\| \le \delta$, $\|\tilde{K}_\alpha\| \le 1$ and (2.10) that

$$\|\tilde{K}_{\alpha_{ME}} A x^\dagger\| \le \|Ax_{\alpha_{ME}}^\delta - y^\delta\| + \|\tilde{K}_{\alpha_{ME}} (Ax^\dagger - y^\delta)\| \le 2\delta\,.$$

Consequently, (2.15) attains for $\alpha = \alpha_{ME}$ the form

$$\|K_{\alpha_{ME}} x^\dagger\| \le (2\delta)^{p/(p+1)} E^{1/(p+1)}\,. \tag{2.16}$$

Let $\alpha = \alpha_*$ be chosen such that the right hand sides of the estimates (2.13) and (2.16) coincide, i.e., let $\alpha_* = (\gamma_{p/2})^{-2/p} E^{-2/(p+1)} (2\delta)^{2/(p+1)}$. If $\alpha_{ME} \le \alpha_*$, then the desired estimate (2.8) follows from the monotonicity property $\|x_{\alpha_{ME}} - x^\dagger\| \le \|x_{\alpha_*} - x^\dagger\|$ (see Theorem 2.3) and from the estimates (2.12), (2.13) with $\alpha = \alpha_*$. If $\alpha_{ME} > \alpha_*$, then the desired estimate (2.8) follows from (2.12) with $\alpha = \alpha_{ME}$ and from the estimates (2.16) and $\gamma_* \delta / \sqrt{\alpha_{ME}} < \gamma_* \delta / \sqrt{\alpha_*}$. Hence, the proof is complete. □

2.4. Newton's iteration for the ME rule in Tikhonov methods

Let us consider the methods of ordinary and iterated Tikhonov regularization (2.5). For choosing the regularization parameter α according to the D principle, the RG rule and the ME rule, respectively, the following nonlinear equations

$$d_D(\alpha) = C\delta\,, \quad d_{RG}(\alpha) = C\delta \quad \text{and} \quad d_{ME}(\alpha) = \delta$$

have to be solved numerically. For the iterative solution of these nonlinear equations the change of the variable α by $r = 1/\alpha$ is reasonable since the functions

$$\tilde{d}_D(r) := d_D(1/r)\,, \quad \tilde{d}_{RG}(r) := d_{RG}(1/r)\,, \quad \tilde{d}_{ME}(r) := d_{ME}(1/r)$$

are *monotonically decreasing and convex* for all $r > 0$. These two properties guarantee global monotone convergence, e.g., for Newton's iteration. For the function $\tilde{d}_{ME}(r)$ we prove the properties of monotonicity and convexity in Theorem 2.5. In order to prove these properties for the functions $\tilde{d}_D(r)$ and $\tilde{d}_{RG}(r)$ we use the representations

$$\tilde{d}_D(r) = \|R_r^m y^\delta\|\,, \quad \tilde{d}_{RG}(r) = \|R_r^{m+1/2} y^\delta\|$$

with $R_r = (I + rAA^*)^{-1}$, exploit the two identities

$$\frac{\mathrm{d}}{\mathrm{d}r}\, \|R_r^k y^\delta\|^2 = -2k\|A^* R_r^{k+1/2} y^\delta\|^2\,, \tag{2.17}$$

$$\frac{\mathrm{d}}{\mathrm{d}r}\, \|A^* R_r^{k+1/2} y^\delta\|^2 = -(2k+1)\|AA^* R_r^{k+1} y^\delta\|^2 \tag{2.18}$$

that hold true for arbitrary $r > 0$ and $k \geq 0$ (compare [35]) and obtain

$$\tilde{d}_D'(r) < 0\,, \quad \tilde{d}_{RG}'(r) < 0\,, \quad \tilde{d}_D''(r) > 0 \quad \text{and} \quad \tilde{d}_{RG}''(r) > 0\,. \tag{2.19}$$

For the proof of (2.19) see also [12] and [4]. In our next theorem we prove that the function

$$\tilde{d}_{ME}(r) = \|R_r^{m+1/2} y^\delta\|^2 / \|R_r^{m+1} y^\delta\|$$

has analogous properties.

Theorem 2.5. *The function $\tilde{d}_{ME}(r)$ is analytic, strictly monotonically decreasing and strictly convex. For all $r > 0$ there hold the estimates*

$$\tilde{d}_{ME}'(r) \leq -m\|A^* R_r^{m+1} y^\delta\|^2 / \|R_r^{m+1} y^\delta\|\,, \tag{2.20}$$

$$\tilde{d}''_{ME}(r) \geq m(m+1)\|R_r^{m+1/2}y^\delta\|^2\|AA^*R_r^{m+2}y^\delta\|^2/\|R_r^{m+1}y^\delta\|^3 . \quad (2.21)$$

Proof. From [35] we have that (2.20) holds true and that $\tilde{d}'_{ME}$ is given by

$$\begin{aligned}\tilde{d}'_{ME}(r) = (m+1)\|R_r^{m+1/2}y^\delta\|^2\|A^*R_r^{m+3/2}y^\delta\|^2/\|R_r^{m+1}y^\delta\|^3 \\ -(2m+1)\|R_r^{m+1}y^\delta\|^2\|A^*R_r^{m+1}y^\delta\|^2/\|R_r^{m+1}y^\delta\|^3 . \quad (2.22)\end{aligned}$$

In order to prove (2.21) we differentiate (2.22) as follows. We use (2.17), (2.18) with $k = m+1/2$ and $k = m+1$, respectively, apply the identity

$$\begin{aligned}\frac{\mathrm{d}}{\mathrm{d}r}\|R_r^{m+1}y^\delta\|^3 &= \frac{\mathrm{d}}{\mathrm{d}r}(\|R_r^{m+1}y^\delta\|^2)^{3/2} \\ &= (3/2)\|R_r^{m+1}y^\delta\|(-2m-2)\|A^*R^{m+3/2}y^\delta\|^2 ,\end{aligned}$$

use the notations

$$\begin{aligned}t_1 &:= \|R_r^{m+1/2}y^\delta\|^2\|A^*R_r^{m+3/2}y^\delta\|^4 \\ t_2 &:= \|A^*R_r^{m+1}y^\delta\|^2\|A^*R_r^{m+3/2}y^\delta\|^2\|R_r^{m+1}y^\delta\|^2 \\ t_3 &:= \|R_r^{m+1}y^\delta\|^4\|AA^*R_r^{m+3/2}y^\delta\|^2 \\ t_4 &:= \|R_r^{m+1/2}y^\delta\|^2\|R_r^{m+1}y^\delta\|^2\|AA^*R_r^{m+2}y^\delta\|^2\end{aligned}$$

and obtain

$$\begin{aligned}&\tilde{d}''_{ME}(r)\|R_r^{m+1}y^\delta\|^5/(m+1) \\ &\quad = 3(m+1)t_1 - 2(2m+1)t_2 + 2(2m+1)t_3 - (2m+3)t_4 \\ &\quad = 3(m+1)(t_1 - t_2 + t_3 - t_4) + (m-1)(t_3 - t_2) + mt_4 . \quad (2.23)\end{aligned}$$

We use the abbreviation $D := (AA^*)^{1/2}$, apply the identity

$$\|z\|^2 = \|R_r^{1/2}z\|^2 + r\|DR_r^{1/2}z\|^2 \quad (2.24)$$

with $z = R_r^{m+1}y^\delta$ and $z = DR_r^{m+1}y^\delta$, respectively, and obtain

$$\begin{aligned}t_3 - t_2 = \|R_r^{m+1}y^\delta\|^2(\|R_r^{m+3/2}y^\delta\|^2\|AA^*R^{m+3/2}y^\delta\|^2 \\ - \|A^*R_r^{m+3/2}y^\delta\|^4) \geq 0 . \quad (2.25)\end{aligned}$$

Repeated use of (2.24) yields

$$\|D^iR_r^{m+2-k/2}y^\delta\|^2 = \sum_{j=0}^{k}\binom{k}{j}r^j a_{i+j} \qquad \text{with} \quad a_j = \|D^jR_r^{m+2}y^\delta\|^2 .$$

We apply this formula for $k = 1, 2, 3$ and $i = 0, \ldots, 3-k$ and obtain

$$\begin{aligned} t_1 - t_2 + t_3 - t_4 &= (a_0 + 3ra_1 + 3r^2 a_2 + r^3 a_3)(a_1 + ra_2)^2 \\ &\quad -(a_0 + 2ra_1 + r^2 a_2)(a_1 + 2ra_2 + r^2 a_3)(a_1 + ra_2) \\ &\quad +(a_0 + 2ra_1 + r^2 a_2)^2 (a_2 + ra_3) \\ &\quad -a_2(a_0 + 2ra_1 + r^2 a_2)(a_0 + 3ra_1 + 3r^2 a_2 + r^3 a_3) \\ &= (r^4 a_2 + 3r^3 a_1 + 3r^2 a_0)(a_3 a_1 - a_2^2) \\ &\quad + r(a_1^{3/2} - a_3^{1/2} a_0)^2 + 2ra_0 a_1[(a_3 a_1)^{1/2} - a_2]. \qquad (2.26) \end{aligned}$$

From the Cauchy—Schwarz inequality we have

$$a_3 a_1 - a_2^2 = \|D^3 R_r^{m+2} y^\delta\|^2 \|DR_r^{m+2} y^\delta\|^2 - (D^3 R_r^{m+2} y^\delta, DR_r^{m+2} y^\delta)^2 \geq 0,$$

and due to (2.26) we obtain $t_1 - t_2 + t_3 - t_4 \geq 0$. From this inequality, (2.25) and (2.23) we conclude that $\tilde{d}''_{ME}(r)\|R_r^{m+1} y^\delta\|^5 \geq m(m+1)t_4$ which is equivalent to (2.21). □

Let us rewrite the derivative (2.22) into some equivalent form which is suitable for numerical computations. We apply the identity $r_{\alpha,k} = y^\delta - Ax^\delta_{\alpha,k} = K_\alpha^k y^\delta = R_r^k y^\delta$ with $K_\alpha = \alpha(AA^* + \alpha I)^{-1}$ and obtain from (2.22)

$$\begin{aligned} \tilde{d}'_{ME}(r) = \big[&(m+1)(r_{\alpha,m}, r_{\alpha,m+1})(A^* r_{\alpha,m+1}, A^* r_{\alpha,m+2}) \\ &- (2m+1)\|A^* r_{\alpha,m+1}\|^2 \|r_{\alpha,m+1}\|^2\big] / \|r_{\alpha,m+1}\|^3 \end{aligned}$$

with $\alpha = 1/r$. This representation shows that the evaluation of the derivative $\tilde{d}'_{ME}(r)$ requires the computation of $r_{\alpha,m}$, $r_{\alpha,m+1}$ and $r_{\alpha,m+2}$. Note that an efficient evaluation of the derivatives $\tilde{d}'_D(r)$ and $\tilde{d}'_{RG}(r)$ requires only the computation of $r_{\alpha,m}$ and $r_{\alpha,m+1}$ (see [4]). As an alternative to Newton's method applied to the equation $\tilde{d}_{ME}(r) - \delta = 0$ one could use the secant method in which the expensive evaluation of $\tilde{d}'_{ME}(r)$ is not necessary.

3. ITERATIVE REGULARIZATION METHODS

3.1. Iteration methods and the ME rule

For approximately solving linear ill-posed problems (1.1) with noisy data $y^\delta \in Y$ we consider iteration methods of the general form

$$x_n^\delta = x_{n-1}^\delta + A^* z_{n-1}, \quad n = 1, 2, \ldots \qquad (3.1)$$

with $z_n \in X$ and initial guess $x_0^\delta = 0$. The elements z_n characterize the special iteration method. For example, $z_n = \beta(y^\delta - Ax_n^\delta)$ with $\beta \in (0, 2/\|A\|^2]$ leads to the well-known Landweber iteration.

Iteration methods for approximately solving ill-posed problems are especially attractive for *large scale problems* (cf. [19]). Such problems arise e.g. in the field of parameter identification in differential equations. Exploiting ideas from control theory, in such identification problems the elements A^*z_n can effectively be computed by solving one direct problem and one associated adjoint problem.

The iterates x_n^δ of the iteration process (3.1) generally diverge. Nevertheless these iterates allow a stable approximation of $x^\dagger$ provided the iteration is stopped after an appropriate number of iteration steps. Two well-known *a posteriori* rules of choosing the stopping index $n = n(\delta)$ are:

1. *Morozov's discrepancy principle* [27, 38, 39]. By this principle (*D principle*) the stopping index in (3.1) is chosen as the first index $n = n_D$ satisfying

$$d_D(n) := \|r_n\| \leq C\delta \quad \text{with} \quad r_n = y^\delta - Ax_n^\delta \quad \text{and} \quad C \geq 1\,. \tag{3.2}$$

2. *Rule of Engl and Gfrerer* [3]. This rule may be applied to iteration methods (3.1) where z_n has the special form $z_n = h_n(\beta_n, AA^*)r_n$ with $r_n = y^\delta - Ax_n^\delta$. Here h_n is some operator function which depends on $\beta_n \in \mathbb{R}$ and β_n is allowed to depend on the noisy data y^δ. In this rule (which we call *EG rule*) the stopping index in (3.1) is chosen as the first index $n = n_{EG}$ satisfying

$$d_{EG}(n) := \frac{(r_n + r_{n+1}, h_n(\beta_n, AA^*)r_n)^{1/2}}{(2\kappa_n)^{1/2}} \leq C\delta \quad \text{with} \quad C \geq 1\,. \tag{3.3}$$

Here κ_n is a constant with

$$\kappa_n = \sup\{h_n(\beta_n, \lambda) \,|\, 0 \leq \lambda \leq \|A\|^2\}. \tag{3.4}$$

In the monotone error rule (ME rule) for choosing the stopping index we focus our attention on the monotonicity property (1.3). Our aim consists in searching for a *largest computable iteration number* n_{ME} for which (1.3) can be guaranteed. Exploiting (3.1) and using the notation $r_n = y^\delta - Ax_n^\delta$

we obtain

$$\begin{aligned}\|x_n^\delta - x^\dagger\|^2 - \|x_{n-1}^\delta - x^\dagger\|^2 &= \|x_{n-1}^\delta + A^* z_{n-1} - x^\dagger\|^2 - \|x_{n-1}^\delta - x^\dagger\|^2 \\ &= 2(x_{n-1}^\delta - x^\dagger, A^* z_{n-1}) + \|A^* z_{n-1}\|^2 = (x_{n-1}^\delta + x_n^\delta - 2x^\dagger, A^* z_{n-1}) \\ &= (2(y^\delta - y) - (r_{n-1} + r_n), z_{n-1}) \\ &\leq 2\|z_{n-1}\|[\delta - (r_{n-1} + r_n, z_{n-1})/(2\|z_{n-1}\|)]\,. \end{aligned} \tag{3.5}$$

This estimate leads us to the following ME rule for iteration methods (3.1):

ME rule. Choose n_{ME} as the first index n satisfying

$$d_{ME}(n) := (r_n + r_{n+1}, z_n)/(2\|z_n\|) \leq \delta\,. \tag{3.6}$$

By this *a posteriori* choice of the stopping index in iteration methods (3.1) the monotonicity property (1.3) can be guaranteed:

Proposition 3.1. *Let* $\|z_n\| \neq 0$ *for* $n = 0, 1, 2, \ldots$ *and let* n_{ME} *be chosen by the ME rule* (3.6). *Then the monotonicity property* (1.3) *holds true.*

Proof. From the ME rule for iteration methods we have that $d_{ME}(n-1) > \delta$ for $n = 1, 2, \ldots, n_{ME}$. Consequently, from (3.5) there follows for all $n = 1, 2, \ldots, n_{ME}$ that

$$\|x_n^\delta - x^\dagger\|^2 \leq \|x_{n-1}^\delta - x^\dagger\|^2 + 2\|z_{n-1}\|\{\delta - d_{ME}(n-1)\} < \|x_{n-1}^\delta - x^\dagger\|^2 \tag{3.7}$$

which completes the proof. □

3.2. The ME rule in gradient type methods

Let us consider gradient type methods of the form (3.1) with $z_n = \beta_n r_n$ where $r_n = y^\delta - Ax_n^\delta$ is the discrepancy and $\beta_n > 0$ is some properly chosen stepsize. For such methods the iteration (3.1) attains the form

$$x_n^\delta = x_{n-1}^\delta + \beta_{n-1} A^*(y^\delta - Ax_{n-1}^\delta)\,, \quad n = 1, 2, \ldots\,. \tag{3.8}$$

Since in the EG rule $h_n(\beta_n, \lambda) = \beta_n$, for the constant κ_n of (3.4) we have $\kappa_n = \beta_n$. Consequently, the functions $d_{EG}(n)$ and $d_{ME}(n)$ of the EG- and ME rules, respectively, attain the form

$$d_{EG}(n) = \frac{(r_n + r_{n+1}, r_n)^{1/2}}{\sqrt{2}} \quad \text{and} \quad d_{ME}(n) = \frac{(r_n + r_{n+1}, r_n)}{2\|r_n\|}\,. \tag{3.9}$$

For gradient type methods (3.8) following properties are valid:

Theorem 3.2. *Let P denote the orthoprojection of Y onto $N(A^*) = \overline{R(A)}^{\perp}$, let $A^*y^\delta \neq 0$ and let β_n be chosen such that*

$$0 < \beta_n \leq \|A^*r_n\|^2/\|AA^*r_n\|^2 . \tag{3.10}$$

Then for the iterates of (3.8) following properties are valid:

(i) The function $d_D(n) = \|r_n\|$ is strictly monotonically decreasing and obeys

$$\|r_{n+1}\|^2 \leq (r_n, r_{n+1}) < \|r_n\|^2 .$$

(ii) The functions $d_{ME}(n)$ and $d_{EG}(n)$ are strictly monotonically decreasing and obey

$$d_D(n+1) < d_{ME}(n) < d_{EG}(n) < d_D(n) .$$

(iii) Let $\beta_n \geq c > 0$ with some positive constant c, then

$$\lim_{n\to\infty} d_{ME}(n) = \lim_{n\to\infty} d_{EG}(n) = \lim_{n\to\infty} d_D(n) = \|Py^\delta\| . \tag{3.11}$$

(iv) If $\|Py^\delta\| < C\delta$, then the stopping indices n_D and n_{EG} are well defined. For $C = 1$ also n_{ME} is well defined and there holds

$$n_D - 1 \leq n_{ME} \leq n_{EG} \leq n_D .$$

(v) If $\|Ax_n^\delta - y^\delta\| \geq \delta$, then

$$\|x_n^\delta - x^\dagger\| < \|x_{n-1}^\delta - x^\dagger\| .$$

Proof. From (3.8) we conclude that $r_{n+1} = (I - \beta_n AA^*)r_n$. Consequently,

$$\text{(a)} \qquad (r_n, r_{n+1}) = \|r_n\|^2 - \beta_n\|A^*r_n\|^2 ,$$

$$\text{(b)} \qquad \|r_{n+1}\|^2 = \|r_n\|^2 - 2\beta_n\|A^*r_n\|^2 + \beta_n^2\|AA^*r_n\|^2 .$$

From (a), $\beta_n > 0$ and $A^*y^\delta \neq 0$ we obtain the right inequality of (i). Combining (a) and (b) we have

$$\|r_{n+1}\|^2 = (r_n, r_{n+1}) - \beta_n\|A^*r_n\|^2 + \beta_n^2\|AA^*r_n\|^2 .$$

From this equation and assumption (3.10) we obtain that the left inequality of (i) holds true. In order to prove the left inequality of (ii) we use the inequality $2ab < a^2 + b^2$ with $a \neq b$ and obtain with the help of the left inequality of (i) that

$$2\|r_n\|\|r_{n+1}\| < \|r_n\|^2 + \|r_{n+1}\|^2 \leq \|r_n\|^2 + (r_n, r_{n+1}),$$

which is equivalent to $d_D(n+1) < d_{ME}(n)$. The middle and right inequalities of (ii) are equivalent to $(r_n, r_{n+1}) < \|r_n\|^2$ and follow from part (i). For proving (3.11) we proceed according to (3.5) and obtain for arbitrary $w \in X$ and arbitrary iteration methods (3.8) with $\beta_n > 0$ that

$$\begin{aligned} \|x_n^\delta - w\|^2 - \|x_{n-1}^\delta - w\|^2 &= (2(y^\delta - Aw) - (r_{n-1} + r_n), \beta_{n-1} r_{n-1}) \\ &\leq 2\beta_{n-1}\|r_{n-1}\|\{\|y^\delta - Aw\| - d_{ME}(n-1)\}. \end{aligned} \quad (3.12)$$

For iteration methods (3.8) with stepsizes β_n satisfying (3.10) there holds

$$\|r_n\|^2 - \|r_{n+1}\|^2 = \beta_n \left(2\|A^*r_n\|^2 - \beta_n\|AA^*r_n\|^2\right) \geq \beta_n\|A^*r_n\|^2.$$

Passing to the limit on both sides yields due to $\beta_n \geq c > 0$ that $\lim_{n\to\infty} \|A^*r_n\| = 0$. From $y^\delta = Ax_n^\delta + r_n$ and the Cauchy—Schwarz-inequality we have

$$\|r_n\|^2 - \|Aw - y^\delta\|^2 = 2(r_n, Aw - Ax_n^\delta) - \|Ax_n^\delta - Aw\|^2 \leq 2\|A^*r_n\|\|x_n^\delta - w\|. \quad (3.13)$$

Now we will use a contradicition argument and assume that $\lim_{n\to\infty} d_D(n) > \|Py^\delta\|$. Under this condition there exists some element $w \in X$ with the property that

$$\lim_{n\to\infty} \|r_n\| > \|Aw - y^\delta\|. \quad (3.14)$$

From (3.14) and (i) we have $\|r_n\| > \|Aw - y^\delta\|$ for all $n \in \mathbb{N}$. This property provides together with (3.12) and (ii) that

$$\|x_n^\delta - w\|^2 - \|x_{n-1}^\delta - w\|^2 < 2\beta_{n-1}\|r_{n-1}\| \{\|r_n\| - d_{ME}(n-1)\} \leq 0.$$

Hence, $\|x_n^\delta - w\|$ is bounded by $\|w\|$ for all $n \in \mathbb{N}$. Passing to the limit on both sides of (3.13) provides due to $\|x_n^\delta - w\| \leq \|w\|$ and $\lim_{n\to\infty} \|A^*r_n\| = 0$ that $\lim_{n\to\infty} \|r_n\| \leq \|Aw - y^\delta\|$. This contradicts (3.14) and shows that our assumption $\lim_{n\to\infty} d_D(n) > \|Py^\delta\|$ cannot hold true. Hence, due to (ii) we obtain (3.11). Assertion (iv) follows from (3.11), the definition of the stopping rules and (ii). In order to prove (v) we use the first part of (3.7), property (ii) as well as the assumption $d_D(n) \geq \delta$ and obtain

$$\begin{aligned} \|x_n^\delta - x^\dagger\|^2 - \|x_{n-1}^\delta - x^\dagger\|^2 &\leq 2\|z_{n-1}\|\{\delta - d_{ME}(n-1)\} \\ &< 2\|z_{n-1}\|\{\delta - d_D(n)\} \leq 0, \end{aligned} \quad (3.15)$$

which finishes the proof. □

Remark 3.1. The limit relation (3.11) can not only be guaranteed for strictly positive stepsizes $\beta_n \geq c > 0$, but also for stepsizes tending to zero not too fast and satisfying

$$0 \leq \beta_i \leq \frac{1}{\|A\|^2} \quad \text{and} \quad \lim_{n\to\infty} \sum_{i=0}^{n} \beta_i = \infty\,.$$

For the proof of this result we use the identity $r_{n+1} = (I - \beta_n AA^*) r_n$ which gives

$$r_{n+1} = r_n(AA^*) y^\delta \quad \text{with} \quad r_n(\lambda) = \prod_{i=0}^{n} (1 - \beta_i \lambda)\,.$$

Hence, for $n \to \infty$ we have for all $\lambda \in (0, \|A\|^2]$

$$|r_n(\lambda)| \to 0 \quad \Leftrightarrow \quad \ln \prod_{i=0}^{n} |1 - \beta_i\lambda| \to -\infty \quad \Leftrightarrow \quad \sum_{i=0}^{n} \ln|1 - \beta_i\lambda| \to -\infty\,.$$

Using the estimate $\ln|1-\xi| \leq -\xi$ for $\xi \in [0,1]$ with $\xi = \beta_i\lambda$ we obtain the result that for $0 \leq \beta_i \leq 1/\|A\|^2$ there holds $|r_n(\lambda)| \to 0$ provided $\sum_{i=0}^{n} \beta_i \to \infty$. Finally we conclude that from $|r_n(\lambda)| \to 0$ for all $\lambda \in (0, \|A\|^2]$ there follows $\lim_{n\to\infty} \|r_n\| = \|Py^\delta\|$ and the proof is complete.

Remark 3.2. Part (v) of Theorem 3.2 shows that the iteration (3.8) should not be stopped as long as $\|Ax_n^\delta - y^\delta\| \geq \delta$ holds. Let us modify the D principle as follows: Choose $n = n_D$ as the first index n satisfying

$$\|Ax_{n+1}^\delta - y^\delta\| < C\delta \quad \text{with} \quad C \geq 1\,.$$

Let n_1 and n_2 the stopping indices of this modified D principle with $C = C_1$ and $C = C_2$, respectively. If $1 \leq C_1 \leq C_2$, then, since $d_D(\alpha)$ is monotonically decreasing there follows $n_1 \geq n_2$, and due to part (v) of Theorem 3.2 we obtain $\|x_{n_1}^\delta - x^\dagger\| \leq \|x_{n_2}^\delta - x^\dagger\|$. Hence, the best possible choice for the constant C in this modified D principle is $C = 1$.

Before we will study some special methods that fit into the framework of Theorem 3.2 let us derive some useful inequality that is helpful for checking condition (3.10).

Proposition 3.3. *Let $D \in \mathcal{L}(Y,Y)$, $D = D^* \geq 0$ and $\eta \geq 0$. Then for all $v \in Y$*

$$\|D^\eta v\| \|Dv\| \leq \|D^{\eta+1} v\| \|v\|\,. \tag{3.16}$$

Proof. We apply the moment inequality (2.14), first with $s = \eta$ and $t = \eta + 1$, second with $s = 1$ and $t = \eta + 1$ and obtain the inequalities

$$\|D^{\eta} v\| \leq \|D^{\eta+1} v\|^{\eta/(\eta+1)} \|v\|^{1/(\eta+1)} \text{ and } \|Dv\| \leq \|D^{\eta+1} v\|^{1/(\eta+1)} \|v\|^{\eta/(\eta+1)}.$$

We multiply both inequalities and obtain (3.16). □

Now we are ready to study some special gradient type methods (3.8) that fit into the framework of Theorem 3.2. In the methods M3 – M5 below we assume that y^{δ} is not an eigenelement of the operator AA^* since in the opposite case there follows $r_1 = 0$ and the stepsizes β_n, $n = 1, 2, \ldots$ are not defined.

Method M1: Landweber's method with $\beta_n = \beta \in (0, 1/\|A\|^2]$. This method may actually be applied with stepsizes $\beta_n = \beta \in (0, 2/\|A\|^2)$, see [1, 4, 5, 13, 17, 23, 25, 28, 30, 38, 39]. However, due to the inequality $\|AA^* r_n\| \leq \|A\| \|A^* r_n\|$ we realize that condition (3.10) and hence the results of Theorem 3.2 hold true for $\beta \in (0, 1/\|A\|^2]$.

Method M2: Nonstationary Landweber's method. This method is characterized by (3.8) with variable $\beta_n \in [c, 1/\|A\|^2]$ and $c > 0$, see [32]. As in method M1 we conclude that condition (3.10) and hence the results of Theorem 3.2 hold true for $\beta_n \in [c, 1/\|A\|^2]$.

Method M3: Steepest descent method (see [1, 2, 7, 9, 10, 11, 16, 20, 22, 31, 32]). This method is characterized by (3.8) with $\beta_n = \|A^* r_n\|^2 / \|AA^* r_n\|^2$. Since this stepsize satisfies (3.10), the results of Theorem 3.2 hold true for this method.

Method M4: α-processes with stepsizes $\beta_n = \|D^{\alpha+1} r_n\|^2 / \|D^{\alpha+2} r_n\|^2$, $D = (AA^*)^{1/2}$ and $\alpha \geq 0$. These methods may actually be applied with $\alpha \geq -1$, see [10, 16, 22, 31, 32]. However, assumption (3.10) holds true only for $\alpha \geq 0$. In order to check (3.10) we apply inequality (3.16) with $D = (AA^*)^{1/2}$, $v = (AA^*)^{1/2} r_n$ and $\eta = \alpha$ and obtain the inequality $\|D^{\alpha+1} r_n\| \|D^2 r_n\| \leq \|D^{\alpha+2} r_n\| \|D r_n\|$ which gives $\beta_n \leq \|D r_n\| / \|D^2 r_n\|$. Since furthermore $\beta_n \geq \|D^{\alpha+1} r_n\|^2 / (\|D\|^2 \|D^{\alpha+1} r_n\|^2) = 1/\|A\|^2$ we obtain

$$1/\|A\|^2 \leq \beta_n \leq \|A^* r_n\|^2 / \|AA^* r_n\|^2 \tag{3.17}$$

and hence (3.10). Since the application of (3.16) requires $\eta \geq 0$ we realize that Theorem 3.2 holds true for α-processes with $\alpha \geq 0$.

Method M5: Method (3.8) with $\beta_n = \max\{\|D^{\alpha+2} r_n\|^2 / (\|D\|^4 \times \|D^{\alpha+1} r_n\|^2), \beta\}$ where $D = (AA^*)^{1/2}$, $\beta \in (0, 1/\|A\|^2]$ and $\alpha \geq -1$ (see

[16, 32]). For this method there holds

$$\beta \le \beta_n \le \max\left\{\frac{\|D\|^2\|D^{\alpha+1}r_n\|^2}{\|D\|^4\|D^{\alpha+1}r_n\|^2}, \beta\right\} = \frac{1}{\|A\|^2}.$$

From this estimate it follows as in method M1 that condition (3.10) and therefore the results of Theorem 3.2 hold true.

Note that method M3 is a special case of α-processes with $\alpha = 0$. In this method the stepsize β_n minimizes the functional $g(\beta_n) := \|r_{n+1}\|^2 = \|(I - \beta_n AA^*)r_n\|^2$.

The results of Theorem 3.2 can be used to establish convergence- and convergence rate results for the above discussed *a posteriori* stopping rules in iteration methods M1–M5.

Theorem 3.4. *Assume $A^*y^\delta \neq 0$ and $\|Py^\delta\| < \delta$. Let x_n^δ the regularized approximation obtained by one of the methods M1 – M5 and let n_D, n_{EG} and n_{ME} the stopping indices according to the discrepancy principle with $C \ge 1$, the EG rule (3.3) with $C \ge 1$ and the ME rule (3.6), respectively. Then for all $n \in \{n_D, n_{ME}, n_{EG}\}$ there holds:*

(i) $\|x_n^\delta - x^\dagger\| \to 0$ for $\delta \to 0$.

(ii) If $x^\dagger \in R((A^*A)^{p/2})$, then

$$\|x_n^\delta - x^\dagger\| = O(\delta^{p/(p+1)}) \quad \text{for all} \quad p > 0.$$

Proof. First, let us discuss the case $C > 1$. In this case the proof for $n = n_D$ is known from the literature, see [38] for method M1, [31, 32] for methods M2–M5 and [9, 10] for method M3. Let $n_{ME,C}$ the stopping index of the ME rule (3.6) with δ replaced by $C\delta$. From the validity of Theorem 3.4 for $n = n_D$ and part (ii) of Theorem 3.2 we obtain that assertions (i), (ii) are valid for n_{EG} and $n_{ME,C}$ as well. Second, let us consider the case $C = 1$. From the monotonicity of $d_{ME}(n)$ we obtain that $n_{ME,C} \le n_{ME,1}$. Consequently, due to Proposition 3.1 there follows

$$\|x_{n_{ME,1}}^\delta - x^\dagger\| \le \|x_{n_{ME,C}}^\delta - x^\dagger\|.$$

Hence, the assertions (i) and (ii) of the theorem are true for $n = n_{ME,1}$. From this result and part (iv) of Theorem 3.2 we conclude that assertions (i) and (ii) of the theorem are also valid for $n = n_D$ and $n = n_{EG}$. □

Some weaker results compared with the results of Theorems 3.2 and 3.4 can be proved for method M4 with $\alpha \in [-1, 0)$. This class of methods contains for $\alpha = -1$ the *minimal error method* in which the stepsize β_n is given by $\beta_n = \|r_n\|^2/\|A^*r_n\|^2$, see [1, 2, 6, 9, 10, 11, 16, 22, 31, 32]. In the minimal error method the stepsize β_n minimizes the norm $\|x_{n+1}^\delta - A^{-1}y^\delta\|$ provided $A^{-1}y^\delta$ exists. For this method there holds $(r_n, r_{n+1}) = 0$ which shows that the left inequality of part (i) of Theorem 3.2 generally does not hold.

Theorem 3.5. *Let $A^*y^\delta \neq 0$. Then in method M4 with $\alpha \in [-1, 0)$ following properties are valid:*

(i) For all $n \in \mathbb{N}$ there holds

$$d_D(n)/2 \leq d_{EG}(n)/\sqrt{2} \leq d_{ME}(n) < d_{EG}(n) < d_D(n)\,.$$

(ii) Denote by $n_{ME,C}$ the stopping index of the ME rule with δ replaced by $C\delta$. Then, $n_{ME,C}$ is well defined for $C > 1$. In addition, $n_{EG,C} := n_{EG}$ is well defined for $C > \sqrt{2}$ and $n_{D,C} := n_D$ is well defined for $C > 2$. For arbitrary $C > 1$ there holds

$$n_{D,2C} \leq n_{EG,\sqrt{2}C} \leq n_{ME,C} \leq n_{EG,C} \leq n_{D,C}\,. \tag{3.18}$$

(iii) For $n = n_{D,2C}$, $n = n_{EG,\sqrt{2}C}$ and $n = n_{ME,C}$ with $C > 1$ there holds

$$\|x_n^\delta - x^\dagger\| \to 0 \quad \textit{for} \quad \delta \to 0\,.$$

Proof. By elementary computations it can be shown that the first two inequalities of assertion (i) are equivalent to $(r_n, r_{n+1}) \geq 0$. This inequality, however, is equivalent to

$$\beta_n \leq \|r_n\|^2/\|A^*r_n\|^2$$

and follows from (3.16) with $\eta = \alpha + 1$ and $v = r_n$. The final two inequalities of assertion (i) are both equivalent to $(r_n, r_{n+1}) < \|r_n\|^2$. This inequality, however, holds for arbitrary stepsize $\beta_n > 0$ since due to $r_{n+1} = [I - \beta_n AA^*]r_n$ and $A^*y^\delta \neq 0$ we have

$$(r_n, r_{n+1}) = \|r_n\|^2 - \beta_n\|A^*r_n\|^2 < \|r_n\|^2\,.$$

Now let us prove indirectly that for $C > 1$ there exists a finite stopping index $n_{ME,C}$. For this aim we assume that $d_{ME}(n) > C\delta$ for all $n \in \mathbb{N}$. Then, due to (3.5), $\|x_n^\delta - x^\dagger\| < \|x_{n-1}^\delta - x^\dagger\|$, and the limit $\lim_{n\to\infty} \|x_n^\delta - x^\dagger\|$ exists.

We summize the inequalities (3.5) with $z_n = \beta_n r_n$ for all $n \geq 1$ and obtain due to $x_0^\delta = 0$ that

$$\|x^\dagger\|^2 - \lim_{n\to\infty} \|x_n^\delta - x^\dagger\|^2 \geq 2\sum_{n=0}^{\infty} \beta_n \|r_n\| \{d_{ME}(n) - \delta\}.$$

Since the right hand side is finite there follows that $\lim_{n\to\infty} \beta_n \|r_n\| \{d_{ME}(n) - \delta\} = 0$. Due to $\beta_n \geq 1/\|A\|^2$, $\|r_n\| > d_{ME}(n)$ (compare (i)) and the assumption $d_{ME}(n) > C\delta$ there follows

$$\beta_n \|r_n\| \geq \|r_n\|/\|A\|^2 > d_{ME}(n)/\|A\|^2 > C\delta/\|A\|^2.$$

Consequently, $\lim_{n\to\infty} \{d_{ME}(n) - \delta\} = 0$. This contradicts our assumption $d_{ME}(n) > C\delta$ for all $n \in \mathbb{N}$ and proves that there has to exist some finite stopping index $n_{ME,C}$ for $C > 1$. Hence, due to the first two inequalities of (i) there follows that $n_{EG,C}$ is well defined for $C > \sqrt{2}$ and $n_{D,C}$ is well defined for $C > 2$. Now the proof of (3.18) follows from the definition of the corresponding stopping rules and (i). In order to prove (iii) we conclude from $n_{D,2C} \leq n_{EG,\sqrt{2}C} \leq n_{ME,C}$ and Proposition 3.1 that

$$\|x^\delta_{n_{ME,C}} - x^\dagger\| \leq \|x^\delta_{n_{EG,\sqrt{2}C}} - x^\dagger\| \leq \|x^\delta_{n_{D,2C}} - x^\dagger\|.$$

Since assertion (iii) holds true for $n_{D,2C}$ with $C > 1$ (see [10]), we conclude that (iii) holds also true for $n_{EG,\sqrt{2}C}$ and $n_{ME,C}$ with $C > 1$. □

Remark 3.3. The estimate (3.5) which led us to the ME rule can also be exploited for finding stepsizes $\beta_n > 0$ in iteration methods (3.8) which guarantee that x_{n+1}^δ is a better approximation for $x^\dagger$ than x_n^δ. Here the stepsize β_n may not only depend on y^δ, but also on the noise level δ. Exploiting (3.7), (3.6) and $r_{n+1} = (I - \beta_n AA^*)r_n$ we obtain for $\beta_n > 0$

$$\|x_{n+1}^\delta - x^\dagger\|^2 - \|x_n^\delta - x^\dagger\|^2 < 2\beta_n \|r_n\| \Big\{\delta - \|r_n\| + \beta_n \frac{\|A^* r_n\|^2}{2\|r_n\|}\Big\}. \quad (3.19)$$

The right hand side of (3.19) is negative for $\|r_n\| > \delta$ and $0 < \beta_n < (2\|r_n\|/\|A^* r_n\|^2)(\|r_n\| - \delta)$ which shows that for such stepsizes the element x_{n+1}^δ is a better approximation for $x^\dagger$ than x_n^δ. Minimizing the right hand side of (3.19) with respect to β_n yields

$$\beta_n = (\|r_n\|/\|A^* r_n\|^2)(\|r_n\| - \delta).$$

Substituting into (3.19) shows (see also [1], p. 69) that for this stepsize the improvement of the squared error can be estimated by

$$\|x_{n+1}^\delta - x^\dagger\|^2 - \|x_n^\delta - x^\dagger\|^2 < -(\|r_n\|^2/\|A^* r_n\|^2)(\|r_n\| - \delta)^2.$$

Some gradient type methods that do not fit into the class of methods (3.8) are *conjugate gradient methods.* The conjugate gradient method for the normal equations $A^*Ax = A^*y^\delta$ (see [1, 2, 9, 10, 11]) is known as powerful method for the approximate solution of ill-posed problems. The advantage of this method over methods of type (3.8) consists in the fact that the stopping index is generally much smaller. In the standard variant of this method (see [1], p. 51) the regularized approximations x_n^δ have the form

$$x_{n+1}^\delta = x_n^\delta + \beta_n p_n\,, \quad p_n = A^* r_n + \gamma_{n-1} p_{n-1}\,, \quad r_n = y^\delta - Ax_n^\delta\,,$$
$$\beta_n = (A^* r_n, p_n)/\|Ap_n\|^2\,, \quad \gamma_{n-1} = \|A^* r_n\|^2/\|A^* r_{n-1}\|^2\,, \quad p_0 = A^* r_0\,.$$

It can easily be realized that this method fits into the framework of methods (3.1). Hence, the ME rule (3.6) for choosing an appropriate stopping index can be applied. In [1] it is shown that for this method the ME function (3.6) has the form

$$d_{ME}(n) = \frac{\|r_n\|^2 + \|r_{n+1}\|^2}{2} \sum_{i=0}^{n} \frac{1}{\|A^* r_i\|^2} \Big/ \sum_{i=0}^{n} \frac{\|r_i\|}{\|A^* r_i\|^2}.$$

This representation for $d_{ME}(n)$ has been used in [1] to prove that the regularized approximation $x_{n_{ME}}^\delta$ converges to $x^\dagger$ for $\delta \to 0$. Unfortunately, we don't know results concerning convergence rates although such results are known for the D principle (see [4]).

3.3. The ME rule in implicit iteration methods

Let us consider implicit iteration methods of the form (3.1) with $z_n = h_n(\beta_n, AA^*) r_n = (\beta_n I + AA^*)^{-1} r_n$ where $r_n = y^\delta - Ax_n^\delta$ is the discrepancy and $\beta_n > 0$ is some suitably chosen real number. For such methods the iteration (3.1) attains the form

$$\begin{aligned} x_n^\delta &= x_{n-1}^\delta + A^*(AA^* + \beta_{n-1} I)^{-1}(y^\delta - Ax_{n-1}^\delta) \\ &= (A^*A + \beta_{n-1} I)^{-1}(\beta_{n-1} x_{n-1}^\delta + A^* y^\delta)\,, \qquad n = 1, 2, \ldots\,. \end{aligned} \tag{3.20}$$

For methods (3.20) the constant κ_n in (3.4) is given by $\kappa_n = \beta_n^{-1}$. From (3.20) we obtain that $r_n = \beta_{n-1}(AA^* + \beta_{n-1} I)^{-1} r_{n-1}$. Consequently, the element $z_n = (\beta_n I + AA^*)^{-1} r_n$ has the form $z_n = \beta_n^{-1} r_{n+1}$ and the functions $d_{EG}(n)$ and $d_{ME}(n)$ of the EG- and ME rules of Subsection 3.1 attain the form

$$d_{EG}(n) = \frac{(r_n + r_{n+1}, r_{n+1})^{1/2}}{\sqrt{2}} \quad \text{and} \quad d_{ME}(n) = \frac{(r_n + r_{n+1}, r_{n+1})}{2\|r_{n+1}\|}\,, \tag{3.21}$$

respectively. For implicit iteration methods with *arbitrary* positive parameters $\beta_n > 0$ following properties are valid:

Theorem 3.6. *Let $A^* y^\delta \neq 0$ and $\beta_n > 0$ arbitrary. Then for the iterates of (3.20) following properties are valid:*

(i) The function $d_D(n) = \|r_n\|$ is strictly monotonically decreasing and obeys

$$\|r_{n+1}\|^2 < (r_n, r_{n+1}) < \|r_n\|^2 .$$

(ii) The functions $d_{ME}(n)$ and $d_{EG}(n)$ are strictly monotonically decreasing and obey

$$\text{(a)} \quad d_D(n+1) < d_{EG}(n) < d_{ME}(n) < d_D(n)$$

$$\text{(b)} \quad \lim_{n\to\infty} d_{ME}(n) = \lim_{n\to\infty} d_{EG}(n) = \lim_{n\to\infty} d_D(n)$$

(iii) Let $\lim_{n\to\infty} d_D(n) < C\delta$ and $C \geq 1$. Then the stopping indices n_D and n_{ME} are well defined. If $C = 1$, then also n_{ME} is well defined and there holds

$$n_D - 1 \leq n_{EG} \leq n_{ME} \leq n_D .$$

(iv) If $\|Ax_n^\delta - y^\delta\| \geq \delta$, then

$$\|x_n^\delta - x^\dagger\| < \|x_{n-1}^\delta - x^\dagger\| .$$

Proof. From (3.20) we conclude that $r_n = (\beta_n^{-1} AA^* + I) r_{n+1}$. Consequently,

$$\text{(a)} \quad (r_n, r_{n+1}) = \|r_{n+1}\|^2 + \beta_n^{-1} \|A^* r_{n+1}\|^2 ,$$

$$\text{(b)} \quad \|r_n\|^2 = \|r_{n+1}\|^2 + 2\beta_n^{-1} \|A^* r_{n+1}\|^2 + \beta_n^{-2} \|AA^* r_{n+1}\|^2 .$$

From (a), (b) and $A^* y^\delta \neq 0$ we obtain assertion (i). From (3.21), the Cauchy-Schwarz inequality, the triangle inequality and inequality (i) we obtain

$$d_{ME}(n) := \frac{(r_n + r_{n+1}, r_{n+1})}{2\|r_{n+1}\|} \leq \frac{1}{2}(\|r_n\| + \|r_{n+1}\|) < \|r_n\| = d_D(n) .$$

Hence, the right inequality of (ii) holds true. The two other inequalities of (ii) are equivalent to $\|r_{n+1}\|^2 < (r_n, r_{n+1})$ and follow from part (i). Now part (b) of (ii) is a consequence of part (a) of (ii) and assertion (iii) of the theorem is a consequence of assertion (ii). The proof of (iv) can be done in analogy to the proof of part (iv) of Theorem 3.2. □

As in explicit iteration methods, property (iv) of the theorem shows that also in implicit iteration methods (3.20) with arbitrary $\beta_n > 0$ the iteration should not be stopped as long as $\|Ax_n^\delta - y^\delta\| \geq \delta$ holds. Here the same discussion as in Remark 3.2 can be made. Further note that the limit $\lim_{n\to\infty} d_D(n)$ always exists since $d_D(n)$ is monotonically decreasing and bounded by zero. However, under the additional condition

$$\lim_{n\to\infty} \sum_{j=0}^{n} \frac{1}{\beta_j} = \infty , \tag{3.22}$$

which is necessary for convergence $x_n \to x^\dagger$ for $n \to \infty$ in the case of exact data, it can be shown that $\lim_{n\to\infty} d_D(n) = \|Py^\delta\|$ where P denotes the orthoprojection of Y onto $N(A^*) = \overline{R(A)}^\perp$ (cf., e.g., [18]). Condition (3.22) excludes that the parameter sequence $\{\beta_n\}$ is growing faster than the sequence $\{n^\alpha\}$ with $\alpha > 1$.

Our next aim in this section is to establish convergence and convergence rate results for implicit iteration methods (3.20) provided the stopping index is chosen from the D principle, the EG rule and the ME rule, respectively. For the D principle there is known that the proof of convergence rates can be done under the additional assumption

$$\frac{1}{\beta_n} \leq c \sum_{j=0}^{n-1} \frac{1}{\beta_j} \quad \text{for some} \quad c > 0 \tag{3.23}$$

(see [18]). These results and the results of Theorem 3.6 allow the proof of convergence and convergence rate results for the EG- and ME rules.

Theorem 3.7. *Assume $A^*y^\delta \neq 0$, $\|Py^\delta\| < \delta$ and (3.22). Let x_n^δ be the regularized approximation obtained by (3.20) and let n_D, n_{EG} and n_{ME} be the stopping indices according to the discrepancy principle with $C \geq 1$, the EG rule (3.3) with $C \geq 1$ and the ME rule (3.6), respectively. Then for all $n \in \{n_D, n_{ME}, n_{EG}\}$ there holds:*

(i) $\|x_n^\delta - x^\dagger\| \to 0$ for $\delta \to 0$.

(ii) If $x^\dagger \in R((A^*A)^{p/2})$ and (3.23) hold, then

$$\|x_n^\delta - x^\dagger\| = O(\delta^{p/(p+1)}) \quad \textit{for all} \quad p > 0 .$$

Proof. The proof is along the lines of the proof of Theorem 3.4 and uses that for $C > 1$ the results of the theorem for $n = n_D$ are known from [18]. □

Now let us study some special implicit iteration methods (3.20) that fit into the framework of Theorem 3.7.

Method M6: Stationary implicit iteration method (see [13, 24, 28, 30, 38, 39]). This method is characterized by (3.20) with fixed $\beta_n := \beta > 0$. Obviously, condition (3.22) holds and condition (3.23) is satisfied with $c = 1$. Hence the results of Theorem 3.7 hold true for this method.

Method M7: Nonstationary implicit iteration method with $\beta_n = \beta q^n$, $\beta > 0$ and $q \in (0,1)$. For this method condition (3.22) can easily be checked and condition (3.23) holds true with $c = 1/q$ (see [18]). Hence, Theorem 3.7 applies for this method.

Method M8: Nonstationary implicit iteration method with $\beta_n \in [c_1, c_2]$, $0 < c_1 < c_2$. For this method (see [32]) property (3.22) is valid due to $\beta_n \le c_2$. From

$$\frac{1}{\beta_n} \le \frac{1}{c_1} \le \frac{1}{c_1} \cdot \frac{c_2}{n} \sum_{j=0}^{n-1} \frac{1}{\beta_j}$$

we obtain (3.23) with $c = c_2/c_1$. Consequently, Theorem 3.7 applies for this method.

Method M9: Nonstationary implicit iteration method with

$$\beta_n = \max\{\|D^{\alpha+2} r_n\|^2 / \|D^{\alpha+1} r_n\|^2 \,, \beta\}, \quad \beta > 0\,, \quad \alpha \ge -1$$

and $D = (AA^*)^{1/2}$. A modification of this method with $\alpha = -1$ and without lower bound β has been studied in [29, 33, 37]. Method M9 (see [16, 32, 37]) is a special case of method M8 with

$$\beta \le \beta_n \le \max\left\{\|D\|^2 \,, \beta\right\} . \tag{3.24}$$

The inequalities (3.24) follow from the definition of β_n. Hence, the results of Theorem 3.7 hold true for this method.

4. NUMERICAL EXPERIMENTS

In this section we summarize some of our numerical results for integral equations of the first kind

$$(Ax)(s) = \int_0^1 K(s,t)x(t)\,\mathrm{d}t = y(s)\,, \quad 0 \le s \le 1 \tag{4.1}$$

Table 1. Errors in the method of ordinary Tikhonov regularization

δ	e_D	e_{RG}	e_{ME}	$e_D/\delta^{1/2}$	$e_{RG}/\delta^{3/5}$	$e_{ME}/\delta^{3/5}$
10^{-1}	.05878	.05982	.05355	.1859	.2381	.2132
10^{-2}	.02332	.01905	.01710	.2332	.3019	.2710
10^{-3}	.00896	.00589	.00527	.2833	.3716	.3325
10^{-4}	.00338	.00183	.00163	.3380	.4597	.4094
10^{-5}	.00119	.00055	.00049	.3763	.5500	.4900

in a L^2-space setting with $X = Y = L^2(0,1)$. We discretized problem (4.1) by the collocation method with 100 piecewise constant spline basis functions on a uniform mesh. Instead of y randomly perturbed data y^δ with $\|y - y^\delta\| \leq \delta$ have been used. Our test problem is taken from [26] and corresponds to equation (4.1) with

$$K(s,t) = \begin{cases} 0 & \text{for } t \leq s \\ 1 & \text{for } s < t\,, \end{cases} \qquad y(s) = \frac{1-\cos(\pi s)}{\pi}\,, \qquad x^\dagger(t) = \sin(\pi t)\,.$$

For this test problem we have $x^\dagger \in R((A^*A)^{p/2})$ for all $p \in (0, 3/2)$.

In a first experiment the discretized problem was regularized by the method of ordinary Tikhonov regularization (2.5) with $m = 1$. The regularization parameter α has been chosen according to the ME rule, the RG rule and the D principle, respectively. For C we have used the constant $C = 1$. In Table 1 the errors

$$e_D = \|x^\delta_{\alpha_D} - x^\dagger\|\,, \quad e_{RG} = \|x^\delta_{\alpha_{RG}} - x^\dagger\| \quad \text{and} \quad e_{ME} = \|x^\delta_{\alpha_{ME}} - x^\dagger\|$$

are given. The results in Table 1 verify the theoretical results of the estimate (2.8) of Theorem 2.2 which tells us that for $x^\dagger \in R((A^*A)^{p/2})$ with $p \leq 2$ there holds $e_{ME} = O(\delta^{p/(p+1)})$. Table 1 also illustrates the well known results that for $1 \leq p \leq 2$ there holds $e_{RG} = O(\delta^{p/(p+1)})$ and $e_D = O(\delta^{1/2})$. In addition we observed that always $e_{ME} < e_{RG}$ holds true.

In a second experiment we solved the discretized problem by one explicit and one implicit iteration method, namely by Landweber's method with $\beta = 1/\|A\|^2$ and by the stationary implicit iteration method with $\beta = 1$. Both iterations were stopped with index $n_* + 10$, where n_* is the first n with $\|x^\delta_{n+1} - x^\dagger\| \geq \|x^\delta_n - x^\dagger\|$. In all experiments the final inequality appeared to be true for all $n = n_* + 1, \ldots, n_* + 10$. In Table 2 comparisons of the indices n_{ME}, n_* and the corresponding errors

$$e_{ME} = \|x_{n_{ME}} - x^\dagger\| \quad \text{and} \quad e_* = \|x_{n_*} - x^\dagger\|$$

Table. 2. Indices n_{ME}, n_* and errors e_{ME}, e_* in iteration methods

δ	Landweber's method				Implicit iteration method			
	n_{ME}	n_*	e_{ME}	e_*	n_{ME}	n_*	e_{ME}	e_*
10^{-1}	15	16	.0630	.0621	32	41	.1131	.1060
10^{-2}	38	67	.0453	.0400	103	166	.0406	.0356
10^{-3}	160	332	.0137	.0111	495	750	.0097	.0087
10^{-4}	947	1545	.0033	.0030	2422	3741	.0031	.0029
10^{-5}	4794	8643	.0009	.0008	12951	19667	.0008	.0008

are given. Note that in these and in all other numerical experiments we observed that $n_{ME} \leq n_*$ and $n_{ME} = n_D$ or $n_{ME} = n_D - 1$, which is in agreement with part (iv) of Theorem 3.2 and part (iii) of Theorem 3.6. We made some further experiments in which instead of random perturbations some special perturbed data y^δ have been used. In these experiments we observed essentially smaller quotients $(n_* - n_{ME})/n_*$ compared with those which follow from Table 2. For some further numerical experiments for comparing the ME rule with the RG rule and the D principle in the method of ordinary Tikhonov regularization see [14, 16, 21, 35].

Acknowledgement

The authors acknowledge assistance from Reimo Palm, University of Tartu, for providing a MATLAB-code for the numerical computations. Parts of this work have been done during a visit of the first author at the University of Applied Sciences Zittau/Görlitz (June–July, 2000).

REFERENCES

1. O. M. Alifanov, E. A. Artjukhin, and S. V. Rumjancev, *Extremal Methods for Solving Ill-Posed Problems and its Applications to the Inverse Problems of the Heat Exchange.* Nauka, Moscow, 1988 (in Russian).

2. O. M. Alifanov and S. V. Rumjancev, On the stability of iterative methods for the solution of linear ill-posed problems. *Soviet Math. Dokl.* (1979) **20**, 1133–1136.

3. H. W. Engl and H. Gfrerer, A posteriori parameter choice for general regularization methods for solving linear ill-posed problems. *Appl. Numer. Math.* (1988) **4**, 395–417.

4. H. W. Engl, M. Hanke and A. Neubauer, *Regularization of Inverse Problems.* Kluwer, Dordrecht, 1996.

5. V. M. Fridman, The method of successive approximations for the Fredholm integral equation of the first kind. *Uspekh. Mat. Nauk* (1956) **11**, 233–234 (in Russian).

6. V. M. Fridman, New methods of solving a linear operator equation. *Dokl. Akad. Nauk SSSR* (1959) **128**, 482–484 (in Russian).

7. V. M. Fridman, On the convergence of methods of steepest descent type. *Uspekh. Mat. Nauk* (1962) **17**, 201–208 (in Russian).

8. H. Gfrerer, An a posteriori parameter choice for ordinary and iterated Tikhonov regularization of ill-posed problems leading to optimal convergence rates. *Math. Comp.* (1987) **49**, 507–522.

9. S. F. Gilyazov, *Methods for Solving Linear Ill-Posed Problems.* Moscow Univ., Moscow, 1987 (in Russian).

10. S. F. Gilyazov, *Approximate Solution of Ill-Posed Problems.* Moscow Univ., Moscow, 1995 (in Russian).

11. S. F. Gilyazov and N. L. Goldman, *Regularization of Ill-Posed Problems by Iteration Methods.* Kluwer, Dordrecht, 2000.

12. V. I. Gordonova and V. A. Morozov, Numerical parameter selection algorithms in the regularization methods. *USSR Comp. Math. Math. Phys.* (1974) **13**, 1–9.

13. U. Hämarik, Monotonicity of error and choice of the stopping index in iterative regularization methods. In: *Differential and Integral Equations: Theory and Numerical Analysis.* A. Pedas (Ed). Estonian Math. Society, Tartu, 1999, 15–30.

14. U. Hämarik and T. Raus, On the a posteriori parameter choice in regularization methods. *Proc. Estonian Acad. Sci. Phys. Math.* (1999) **48**, 133–145.

15. U. Hämarik and U. Tautenhahn, The monotone error rule for parameter choice in regularization methods. In: *ENUMATH 99-Proc. of the 3rd European Conf. on Numer. Math. and Advanced Appl.* P. Neittaanmäki, T. Tiihonen, and P. Tarvainen (Eds). World Scientific, Singapore, 2000, 518–525.

16. U. Hämarik and U. Tautenhahn, On the monotone error rule for parameter choice in iterative and continuous regularization methods. *BIT* (2001) **41**, 1029–1038.

17. M. Hanke, Accelerated Landweber iterations for the solution of ill-posed equations. *Numer. Math.* (1991) **60**, 341–373.

18. M. Hanke and C. W. Groetsch, Nonstationary iterated Tikhonov regularization. *J. Optimization Theory Appl.* (1998) **98**, 37–53.

19. M. Hanke and P. C. Hansen, Regularization methods for large-scale problems, *Surv. Math. Ind.* (1993) **4**, 253–315.

20. W. J. Kammerer and M. Z. Nashed, Steepest descent for singular linear operators with nonclosed range. *Appl. Anal.* (1971) **1**, 143–159.

21. E. M. Kiss, *Ein Beitrag zur Regularisierung und Diskretisierung des inversen Problems der Identifikation des Memorykernes in der Viskoelastizität.* Dissertation. Techn. Univ. Bergakademie Freiberg, Shaker, Aachen, 2000.

22. M. A. Krasnoselskij, G. M. Vainikko, P. P. Zabreiko, *et al*, *Approximate Solution of Operator Equations.* Wolters-Noordhoff, Groningen, 1972.

23. L. Landweber, An iteration formula for Fredholm integral equations of the first kind. *Amer. J. Math.* (1951) **73**, 615–624.

24. L. J. Lardy, A series representation for the generalized inverse of a closed linear operator. *Atti della Accademia Nazionale dei Lincei, Rendiconti della Classe di Science Fisiche Matematiche, e Naturali. Serie VIII* (1975) **58**, 152–157.

25. M. M. Lavrent'ev, *Some Improperly Posed Problems of Mathematical Physics.* Springer, New York, 1967.

26. A. K. L. Louis, *Inverse und schlecht gestellte Probleme.* Teubner, Stuttgart, 1989.

27. V. A. Morozov, On the solution of functional equations by the method of regularization. *Soviet Math. Dokl.* (1966) **7**, 414–417.

28. V. A. Morozov, On the regularizing families of operators. *Vychisl. Metody Program* (1967) **8**, 63–95 (in Russian).

29. S. V. Pereverzev and E. Schock, Brakhage's implicit iteration method and information complexity of equations with operators having closed range. *J. Complexity* (1999) **15**, 881–890.

30. T. Raus, On the discrepancy principle for solution of ill-posed problems with non-selfadjoint operators. *Acta et Comment. Univ. Tartuensis* (1985) **715**, 12–20 (in Russian).

31. L. Sarv, A certain family of nonlinear iterative methods for solving ill-posed problems. *Izv. Akad. Nauk Est. SSR. Fiz. Mat.* (1982) **31**, 261–269 (in Russian).

32. L. Sarv, *Iteration Methods for Linear Ill-Posed Problems.* Dissertation. Sverdlovsk, 1984 (in Russian).

33. V. I. Starostenko and S. M. Oganesjan, Regularizing iterative processes. In: *Methods for Solution of Ill-Posed Problems and its Applications.* A. N. Tikhonov (Ed). Novosibirsk, 1982, 119–128 (in Russian).

34. U. Tautenhahn, On a new parameter choice rule for ill-posed inverse problems. In: *Hellenic-European Research on Computer Mathematics and its Applications.* E. A. Lipitakis (Ed). Lea Publishers, Athens, 1999, 118–125.

35. U. Tautenhahn and U. Hämarik, The use of monotonicity for choosing the regularization parameter in ill-posed problems. *Inverse Problems* (1999) **15**, 1487–1505.

36. A. N. Tikhonov and V. Y. Arsenin, *Solution of Ill-Posed Problems.* Wiley, New York, 1977.

37. V. N. Trushnikov, A non-linear regularizing algorithm and some of its applications. *USSR Comp. Math. Math. Phys.* (1979) **19**, 22–31.

38. G. M. Vainikko, The discrepancy principle for a class of regularization methods. *USSR Comp. Math. Math. Phys.* (1982) **22**, 1–19.

39. G. M. Vainikko and A. Y. Veretennikov, *Iteration Procedures in Ill-Posed Problems.* Nauka, Moscow, 1986 (in Russian).

Ill-Posed and Non-Classical Problems of
Mathematical Physics and Analysis, Samarkand, 2000, pp. 57–69
M. M. Lavrent'ev and S. I. Kabanikhin (Eds)

Well-posedness of one-dimensional inverse acoustic problem in L_2 for small depth or small data

S. I. KABANIKHIN*, K. T. ISKAKOV†, and M. YAMAMOTO‡

Abstract — We investigate a system of nonlinear Volterra integral equations. In previous publications [1, 2] we showed that this system can be derived from an inverse acoustic problem and from an inverse electromagnetic problem respectively. Using the method of weighted estimates and the Banach fixed point theorem we prove that the inverse problem is well-posed if either the domain is sufficiently small or the data is sufficiently small. Moreover we find explicit constants which allow us to estimate the depth of well-posedness provided the data are arbitrary but fixed or to estimate the norm of the data which guarantees the existence while the depth is arbitrary but fixed. In [3] L_2-approach has been applied to general nonlinear Volterra operator equation, but without specifications of the constants. In [4] the similar technique allowed to justify the method of steepest descent. Multidimensional analog of our problem was considered in [5].

1. INTRODUCTION

We consider a system of nonlinear Volterra integral equations

$$\Phi(x,t) = G(x,t) + \mathcal{B}(\Phi), \qquad (x,t) \in \Delta(l). \tag{1.1}$$

*Sobolev Institute of Mathematics, Acad. Koptyuga prosp., 4, Novosibirsk, 630090, Russia. E-mail: kabanikh@math.nsc.ru

†Buketov Karaganda State University, Universitetskaya str., 28, Karaganda, 470074, Kazakhstan. E-mail: kazizat@mail.ru

‡Graduate School of Mathematics, University of Tokyo, Komaba, Tokyo, Japan. E-mail: myama@ms.u-tokyo.ac.jp

Here

$$\Phi(x,t) = (\Phi_1(x,t), \Phi_2(x), \Phi_3(x))^\top, \qquad G(x,t) = (G_1(x,t), G_2, G_3(x))^\top,$$
$$G_1 = G_1(x,t), \qquad G_j = G_j(x), \qquad j = 2,3,$$
$$\Phi_1(x,t) = u_x(x,t), \qquad \Phi_2(x) = p(x), \qquad \Phi_3(x) = a(x),$$
$$G_1(x,t) = F_x(x,t), \qquad G_2 = -1/\alpha, \qquad G_3(x) = -2f'(2x)/\alpha,$$
$$F_x(x,t) = [f'(t+x) - f'(t-x)]/2, \quad \Delta(l) = \{(x,t) \mid 0 < x < t < 2l - x\},$$

$$\mathcal{B}_1(\Phi) = \frac{1}{2}\int_0^x \Phi_3(\xi)[\Phi_1(\xi, t+x-\xi) + \Phi_1(\xi, t-x+\xi)]\, d\xi, \tag{1.2}$$
$$\mathcal{B}_2(\Phi) = -\frac{1}{2}\int_0^x \Phi_3(\xi)\Phi_2(\xi)\, d\xi, \tag{1.3}$$
$$\mathcal{B}_3(\Phi) = -\alpha G_3(x)\mathcal{B}_2(\Phi) - 2\mathcal{P}(\Phi)/\alpha + 2\mathcal{B}_2(\Phi)\mathcal{P}(\Phi), \tag{1.4}$$
$$\mathcal{P}(\Phi) = \int_0^x \Phi_3(\xi)\Phi_1(\xi, 2x-\xi)\, d\xi, \tag{1.5}$$

In [1, 2] Volterra integral equation (1.1) is derived from an inverse acoustic problem and an inverse electromagnetic problem respectively, and inverse problem are reduced to (1.1).

We set

$$\|\Phi\|^2(x) = \sum_{j=1}^{3} \|\Phi_j\|^2(x), \qquad x \in (0,l), \tag{1.6}$$

$$\|\Phi_1\|^2_{L_2(\Delta(l,x))} = \int_0^x \int_\xi^{2l-\xi} \Phi_1^2(\xi,\tau)\, d\tau\, d\xi,$$

$$\|\Phi_j\|^2_{L_2(0,x)} = \int_0^x \Phi_j^2(\xi)\, d\xi, \qquad j = 2,3.$$

The space of vector-functions $\Phi(x,t) = (\Phi_1(x,t), \Phi_2(x), \Phi_3(x))^\top$ with the norm (1.6) will be referred as $L_2(l)$. We define the operator

$$V(\Phi) = G(x,t) + \mathcal{B}(\Phi), \quad (x,t) \in \Delta(l). \tag{1.7}$$

Lemma 1.1. *If $G \in L_2(l)$, then*

$$V : \ L_2(l) \longrightarrow L_2(l). \tag{1.8}$$

The proof of the Lemma 1.1 will be given in Appendix 1.

We define k-norms as follows ($k \geq 0$):

$$\|\Phi_1\|_k^2(x) = \int_0^x \|\Phi_1\|_{(1)}^2(\xi)\, \mathrm{e}^{-2k\xi}\, d\xi, \qquad x \in (0,l),$$

$$\|\Phi_1\|_{(1)}^2(x) = \int_x^{2l-x} \Phi_1^2(x,t)\, dt, \qquad x \in (0,l),$$

$$\|\Phi_j\|_k^2(x) = \int_0^x \Phi_j^2(\xi)\, \mathrm{e}^{-2k\xi}\, d\xi, \qquad x \in (0,l), \quad j = 2,3,$$

and the ball for $G \in L_2(l)$, $\varepsilon > 0$, $k \geq 0$

$$\Pi(G,\varepsilon,k) = \{\Phi \in L_2(l) \mid \|G - \Phi\|_k(l) < \varepsilon\},$$

where $\|\Phi\|_k^2(x) = \sum\limits_{j=1}^{3} \|\Phi_j\|_k^2(x)$.

Lemma 1.2. *Let us consider the operator $V(\Phi)$ define by (1.7). Let k satisfy*

$$16l^2 + 2l + 3/2 < k \tag{1.9}$$

and

$$G_*^2 \equiv \|G\|^2(l) < \mathrm{e}^{-2kl}/2, \tag{1.10}$$

$$\varepsilon^2 < \mathrm{e}^{-2kl}/2. \tag{1.11}$$

Then

$$V(\Pi(G,\varepsilon,k)) \subset \Pi(G,\varepsilon,k). \tag{1.12}$$

The proof is given in Appendix 2.

Lemma 1.3. *Let $G \in L_2(l)$ and $\alpha \neq 0$, $l > 0$, $k \in (M_3(l,\alpha),\infty)$,*

$$M_2(l,\alpha) = 24/\alpha^2 + 5l + 4\alpha^2 + 16, \quad M_3(l,\alpha) = \max\{M_2(l,\alpha), 16l^2 + 2l + 3/2\}.$$

If G_* and $\varepsilon > 0$ satisfy (1.10) and (1.11), then V is a contraction operator on $\Pi(G,\varepsilon,k)$ with contraction constant $q^2 = M_3(l,\alpha)/k$.

Proof. Let $\Phi^{(1)}, \Phi^{(2)} \in \Pi(G,\varepsilon,k)$. We denote $\tilde{\Phi} = \Phi^{(1)} - \Phi^{(2)}$ and estimate

$$\Upsilon_{(j)} = V(\Phi^{(1)})_{(j)} - V(\Phi^{(2)})_{(j)}, \qquad j = 1,2,3$$

$$(\Phi_*^2 = \max_{j=1,2} \|\Phi^{(j)}\|^2(l), \qquad G_*^2 = \|G\|^2(l)).$$

Then we have

$$\begin{aligned}\Upsilon_{(1)}(x,t) &= \mathcal{B}_1(\Phi^{(1)}) - \mathcal{B}_1(\Phi^{(2)}) \\ &= \frac{1}{2}\int_0^x \tilde{\Phi}_3(\xi)[\Phi_1^{(1)}(\xi, t+x-\xi) + \Phi_1^{(1)}(\xi, t-x+\xi)]\,d\xi \\ &\qquad + \frac{1}{2}\int_0^x \Phi_3^{(2)}(\xi)[\tilde{\Phi}_1^{(}\xi, t+x-\xi) + \tilde{\Phi}_1(\xi, t-x+\xi)]\,d\xi,\end{aligned}$$

and

$$\begin{aligned}|\Upsilon_{(1)}(x,t)| \le \frac{1}{2}\Big[\int_0^x |\tilde{\Phi}_3(\xi)|^2\, \mathrm{e}^{-2k\xi}\, d\xi\Big]^{1/2} \\ \times \Big[\int_0^x [\Phi_1^{(1)}(\xi, t+x-\xi) + \Phi_1^{(1)}(\xi, t-x+\xi)]^2\, \mathrm{e}^{2k\xi}\, d\xi\Big]^{1/2} \\ + \frac{1}{2}\Big[\int_0^x |\Phi_3^{(2)}(\xi)|^2\, \mathrm{e}^{2k\xi}\, d\xi\Big]^{1/2} \\ \times \Big[\int_0^x [\tilde{\Phi}_1(\xi, t+x-\xi) + \tilde{\Phi}_1(\xi, t-x+\xi)]^2\, \mathrm{e}^{-2k\xi}\, d\xi\Big]^{1/2},\end{aligned}$$

$$\begin{aligned}|\Upsilon_{(1)}(x,t)|^2 \le \frac{1}{2}\,\|\tilde{\Phi}_3\|_k^2(x) \int_0^x [\Phi_1^{(1)}(\xi, t+x-\xi) + \Phi_1^{(1)}(\xi, t-x+\xi)]^2\, \mathrm{e}^{2k\xi}\, d\xi \\ + \frac{1}{2}\,\|\Phi_3^{(2)}\|_{-k}^2(x) \int_0^x [\tilde{\Phi}_1(\xi, t+x-\xi) + \tilde{\Phi}_1(\xi, t-x+\xi)]^2\, \mathrm{e}^{-2k\xi}\, d\xi,\end{aligned}$$

$$\begin{aligned}\|\Upsilon_{(1)}\|_{(1)}^2(x) &\le 2\|\tilde{\Phi}_3\|_k^2(x) \int_0^x \mathrm{e}^{2k\xi}\, \|\Phi_1^{(1)}\|_{(1)}^2(\xi)\, d\xi \\ &\qquad + 2\|\Phi_3^{(2)}\|_{-k}(x) \int_0^x \mathrm{e}^{-2k\xi}\, \|\tilde{\Phi}_1\|_{(1)}^2\, d\xi \\ &\le \frac{1}{k}\,\Phi_*^2\, \mathrm{e}^{2kx}(\|\tilde{\Phi}_3\|_k(x) + \|\tilde{\Phi}_1\|_k(x)), \qquad x \in (0,l). \quad (1.13)\end{aligned}$$

Now we consider $\Upsilon_{(2)}$:

$$\Upsilon_{(2)} = \mathcal{B}_2(\Phi^{(1)}) - \mathcal{B}_2(\Phi^{(2)}) = -\frac{1}{2}\int_0^x \tilde{\Phi}_3(\xi)\Phi_2^{(1)}(\xi)\,d\xi - \frac{1}{2}\int_0^x \Phi_3^{(2)}(\xi)\tilde{\Phi}_2(\xi)\,d\xi.$$

Hence

$$|\Upsilon_{(2)}(x,t)| \le \frac{1}{2}\Big[\int_0^x |\tilde{\varphi}_3(\xi)|^2\,\mathrm{e}^{-2k\xi}\,d\xi\Big]^{1/2}\Big[\int_0^x |\Phi_2^{(1)}(\xi)|^2\,\mathrm{e}^{2k\xi}\,d\xi\Big]^{1/2}$$
$$+\frac{1}{2}\Big[\int_0^x |\Phi_3^{(2)}(\xi)|^2\,\mathrm{e}^{2k\xi}\,d\xi\Big]^{1/2}\Big[\int_0^x |\tilde{\Phi}_2(\xi)|^2\,\mathrm{e}^{-2k\xi}\,d\xi$$
$$\le \frac{1}{2}\,\|\tilde{\Phi}_3\|_k(x)\|\Phi_2^{(1)}\|_{-k}(x) + \frac{1}{2}\,\|\Phi_3^{(2)}\|_{-k}(x)\|\tilde{\Phi}_2\|_k(x)$$

and

$$|\Upsilon_{(2)}(x)|\,\mathrm{e}^{-kx} \le \frac{1}{2}\,\mathrm{e}^{-kx}(\|\tilde{\Phi}_3\|_k(x)\|\Phi_2^{(1)}\|_{-k}(x) + \|\Phi_3^{(2)}\|_{-k}(x)\|\tilde{\Phi}_2\|(x)),$$

$$\|\Upsilon_{(2)}\|_k^2(x) \le \frac{1}{2}\int_0^x \mathrm{e}^{-2k\xi}(\|\tilde{\Phi}_3\|_k(\xi)\|\Phi_2^{(1)}\|_{-k}(\xi) + \|\Phi_3^{(2)}\|^{-k}(\xi)\|\tilde{\Phi}_2\|_k(\xi))\,d\xi$$
$$\le \frac{1}{4k}\,\Phi_*^2\,\mathrm{e}^{2kx}(\|\tilde{\Phi}_2\|_k(x) + \|\tilde{\Phi}_3\|_k(x)), \quad (1.14)$$

Let us estimate $\Upsilon_{(3)}$.

$$\Upsilon_{(3)}^2(x) \le 4\alpha^2|G_3(x)|^2\Upsilon_{(2)}^2(x) + 8Y^2(x)/\alpha^2$$
$$+\,8\Upsilon_{(2)}^2(x)\mathcal{P}^2(\Phi^{(1)}) + 8Y^2(x)\mathcal{B}_2^2(\Phi^{(2)}), \quad (1.15)$$

$$Y^2(x) \le 2\int_0^x |\tilde{\Phi}_3(\xi)|^2\,d\xi \int_0^x |\Phi_1^{(1)}(\xi, 2x-\xi)|^2\,d\xi$$
$$+\,2\int_0^x |\Phi_3^{(2)}(\xi)|^2\,d\xi \int_0^x |\tilde{\Phi}_1(\xi, 2x-\xi)|^2\,d\xi, \quad (1.16)$$

$$\mathcal{P}^2(\Phi^{(1)}) \le \int_0^x |\Phi_3^{(1)}(\xi)|^2\,d\xi \int_0^x |\Phi_1^{(1)}(\xi, 2x-\xi)|^2\,d\xi$$
$$\le \Phi_*^2 \int_0^x |\Phi_1^{(1)}(\xi, 2x-\xi)|^2\,d\xi, \quad (1.17)$$

$$\mathcal{B}_2^2(\Phi^{(2)}) \le \Phi_*^4/4. \quad (1.18)$$

Therefore

$$\begin{aligned}
\Upsilon^2_{(3)}(x) \le \alpha^2 |G_3(x)|^2 \frac{1}{k}\,\Phi^2_*\, \mathrm{e}^{2kx}\left(\|\tilde\Phi_2\|^2_k(x) + \|\tilde\Phi_3\|^2_k(x)\right) \\
+ 2\Big(\frac{8}{\alpha^2} + 2\Phi^4_*\Big)\Big[\int_0^x |\tilde\Phi_3(\xi)|^2\,d\xi \int_0^x |\Phi_1^{(1)}(\xi, 2x-\xi)|^2\,d\xi \\
+ \Phi^2_* \int_0^x |\tilde\Phi_1(\xi, 2x-\xi)|^2\,d\xi\Big] \\
+ \frac{2}{k}\,\Phi^2_*\,\mathrm{e}^{2kx}\left(\|\tilde\Phi_2\|^2_k(x) + \|\tilde\Phi_3\|^2_k(x)\right)\Phi^2_* \int_0^x |\Phi_1^{(1)}(\xi, 2x-\xi)|^2\,d\xi \\
\le \frac{1}{k}\,\mathrm{e}^{2kx}\left[\alpha^2\Phi^2_*|G_{(3)}(x)|^2\left(\|\tilde\Phi_2\|^2_k(x) + \|\tilde\Phi_3\|^2_k(x)\right)\right] \\
+ \frac{1}{k}\,\mathrm{e}^{2kx}\left(\frac{8}{\alpha^2} + 2\Phi^4_*\right)\Big[\|\tilde\Phi_3\|^2_k(x)\int_0^x |\Phi_1^{(1)}(\xi, 2x-\xi)|^2\,d\xi \\
+ \Phi^2_*\int_0^x \mathrm{e}^{-2k\xi}\,|\tilde\Phi_1(\xi, 2x-\xi)|^2\,d\xi\Big] \\
+ \frac{1}{k}\,\mathrm{e}^{2kx}\Big[2\Phi^4_*\left(\|\tilde\Phi_2\|^2_k(x) + \|\tilde\Phi_3\|^2_k(x)\right)\int_0^x |\Phi_1^{(1)}(\xi, 2x-\xi)|^2\Big].
\end{aligned} \tag{1.19}$$

We multiply (1.19) by e^{-2kx} and integrate from 0 to x with respect to x

$$\begin{aligned}
\|\Upsilon_{(3)}\|^2_k(x) \le \frac{1}{k}\int_0^x \Phi^2_*\alpha^2\,|G_{(3)}(\xi)|^2\left[\|\tilde\Phi_2\|^2_k(\xi) + \|\tilde\Phi_3\|^2_k(\xi)\right]d\xi \\
+ \frac{1}{k}\Big(\frac{8}{\alpha^2} + 2\Phi^4_*\Big)\Big[\int_0^x \|\tilde\Phi_3\|^2_k(\xi)\int_0^\xi |\Phi_1^{(1)}(\zeta, 2\tau-\zeta)|^2\,d\zeta\,d\xi + \Phi^2_*\|\tilde\Phi_1\|^2_k(x)\Big] \\
+ \frac{2}{k}\,\Phi^4_*\int_0^x \left[\|\tilde\Phi_2\|^2_k(\xi) + \|\tilde\Phi_3\|^2_k(\xi)\right]\int_0^\xi |\Phi_1^{(1)}(\zeta, 2\xi-\zeta)|^2\,d\zeta\,d\xi.
\end{aligned} \tag{1.20}$$

After summation of (1.13), (1.14), (1.20) we obtain

$$\|\Upsilon(\Phi^{(1)}) - \Upsilon(\Phi^{(2)})\|^2_k(x) \le \frac{1}{k}\Big[\omega_{(0)}\|\tilde\Phi_1\|^2_k(x) + \int_0^x \sum_{j=1}^3 \omega_{(j)}(\xi)\|\tilde\Phi_j\|^2_k(\xi)\,d\xi\Big],$$
$$x \in (0, l). \tag{1.21}$$

Here

$$\omega_{(0)} = (8/\alpha^2 + 2\Phi_*^4)\Phi_*^2, \qquad \omega_{(1)} = \Phi_*^2,$$

$$\omega_{(2)} = \frac{1}{4}\,\Phi_*^2 + \alpha^2 |G_3(\xi)|^2 \Phi_*^2 + 2\Phi_*^4 \int_0^\xi |\Phi_1^{(1)}(\zeta, 2\xi - \zeta)|^2 \, d\zeta,$$

$$\omega_{(3)} = \frac{5}{4}\,\Phi_*^2 + \alpha^2 |G_3(\xi)|^2 \Phi_*^2 + \Big(\frac{8}{\alpha^2} + 2\Phi_*^4\Big) \int_0^\xi |\Phi_1^{(1)}(\zeta, 2\xi - \zeta)|^2 \, d\zeta.$$

Since (1.21) is valid for all $x \in (0, l)$, we can obtain an upper estimate

$$\|\Upsilon(\Phi^{(1)}) - \Upsilon(\Phi^{(2)})\|_k^2(l) \le \frac{1}{k}\,\omega_* \|\Phi^{(1)} - \Phi^{(2)}\|_k^2(l) \tag{1.22}$$

with

$$\omega_* = \omega_{(0)} + \int_0^l \sum_{j=1}^{3} \omega_{(j)}(\xi)\, d\xi.$$

We recall that $\Phi_*^2 = \max_{j=1,2} \|\Phi^{(j)}\|^2(l)$ and $G_* = \|G\|^2(l)$. Therefore

$$\omega_* \le M_1(l, \alpha, \Phi_*, G_*) = \Phi_*^2(12/\alpha^2 + 5l/2 + 2\alpha^2 G_*^2 + 4\Phi_*^4). \tag{1.23}$$

Therefore the condition for the operator V to be a contraction operator is

$$k \in (M_1(l, \alpha, \Phi_*, G_*), \infty). \tag{1.24}$$

If $\Phi \in \Pi(G, \varepsilon, k)$ then

$$\|\Phi\|_k(l) \le \|G\|_k(l) + \varepsilon \le G_* + \varepsilon$$

and hence

$$\Phi_*^2 \le \mathrm{e}^{2kl}\, \|\Phi\|_k^2(l) \le \mathrm{e}^{2kl} \quad (\le G_* + \varepsilon)^2. \tag{1.25}$$

Due to the conditions (1.10), (1.11),

$$(G_* + \varepsilon)^2 \le 2G_*^2 + 2\varepsilon^2 < 2\,\mathrm{e}^{-2kl} \tag{1.26}$$

and hence

$$\Phi_*^2 \le 2, \tag{1.27}$$

$$G_*^2 \le 1. \tag{1.28}$$

Therefore we can estimate

$$M_1(l, \alpha, \Phi_*, G_*) < M_2(l, \alpha) = 2(12/\alpha^2 + 5l/2 + 2\alpha^2 + 8) \tag{1.29}$$

which completes the proof. □

Theorem 1.1 [well-posedness for small data]. *Let $G \in L_2(l)$, $l > 0$, $\alpha \neq 0$,*

$$k \in (M_3(l,\alpha), \infty),$$

$$M_3(l,\alpha) = \max\{24/\alpha^2 + 5l + 4\alpha^2 + 16,\ 16l^2 + 2l + 3/2\}, \tag{1.30}$$

$$\varepsilon^2 \in (0, \mathrm{e}^{-2kl}/2). \tag{1.31}$$

If

$$\|G\|^2(l) < \varepsilon^2, \tag{1.32}$$

then the system (1.1) *has the unique solution $\Phi \in L_2(l)$ which depends continuously on data G.*

Proof. According to Lemma 1.3 and the Banach fixed point theorem, the operator V has the unique fixed point $\bar{\Phi} \in \Pi(G, \varepsilon, k)$ which solves the system

$$\Phi(x,t) = V(\Phi)(x,t), \qquad (x,t) \in \Delta(l). \tag{1.33}$$

Let us prove the stability. Let $\Phi^{(j)}$, $j = 1, 2$ solve the system

$$\Phi^{(j)}(x,t) = G^{(j)}(x,t) + \mathcal{B}(\Phi^{(j)}). \tag{1.34}$$

We denote

$$\tilde{\Phi}(x,t) = \Phi^{(1)}(x,t) - \Phi^{(2)}(x,t), \qquad \tilde{G}(x,t) = G^{(1)}(x,t) - G^{(2)}(x,t)$$

and obtain after the subtraction

$$\tilde{\Phi}(x,t) = \tilde{G}(x,t) + \mathcal{B}(\Phi^{(1)}) - \mathcal{B}(\Phi^{(2)}). \tag{1.35}$$

According to Lemma 1.3

$$\|\tilde{\Phi}\|_k(l) \leq \|\tilde{G}\|_k(l) + q\|\tilde{\Phi}\|_k(l),$$

and, since $q \in (0,1)$, we obtain

$$\|\tilde{\Phi}\|_k(l) \leq \frac{1}{1-q}\|\tilde{G}\|_k(l) \leq \frac{1}{1-q}\|\tilde{G}\|. \tag{1.36}$$

Using the property of the k-norm:

$$\|\tilde{\Phi}\|(l) \leq \mathrm{e}^{2kl}\|\tilde{\Phi}\|_k(l),$$

we obtain from (1.36)

$$\|\tilde{\Phi}\|(l) \leq \frac{1}{1-q}\mathrm{e}^{2kl}\|\tilde{G}\|(l). \tag{1.37}$$

□

2. LOCAL WELL-POSEDNESS

Theorem 2.1 [local well-posedness]. *Let $G \in L_2(l_0)$, $l_0 > 0$, $\alpha \neq 0$. Then there exists $l_* \in (0, l_0)$ such that for every $l \in (0, l_*)$ the solution Φ of the system (1.1) exists and is unique in $L_2(l)$ and depends continuously on data $\tilde{G} \in L_2(l)$ ($\tilde{G}$ is the restriction of G on $\Delta(l)$).*

Proof. For $l \in (0, l_0)$ we define the operator V by (1.7). Due to Lemma 1.1, we see that $V : L_2(l) \longrightarrow L_2(l)$. Let us examine estimate (3.11) obtained in Appendix 2:

$$\|V(\Phi) - G\|_k^2 \leq \frac{l\Phi_*^2}{k}\left[\frac{5}{2} + \frac{3}{8l}\|G\|^2(l) + \frac{6l}{\alpha^2} + l\Phi_*^2\right]\Phi_k^2. \tag{2.1}$$

We suppose that $\Phi \in \Pi(G, \varepsilon, k)$, therefore

$$\Phi_k^2 \leq (G_k + \varepsilon)^2 \leq 2G_k^2 + 2\varepsilon^2 \leq 2G_*^2 + 2\varepsilon^2, \tag{2.2}$$

$$\Phi_*^2 \leq \mathrm{e}^{2kl}\,\Phi_k^2 \leq 2\,\mathrm{e}^{2kl}(G_*^2 + \varepsilon^2). \tag{2.3}$$

Then it follows from (2.1)

$$\|V(\Phi) - G\|_k^2 \leq \frac{1}{k}\,\mathrm{e}^{2kl}\,4(G_*^2 + \varepsilon^2) \times \left[\frac{5l}{2} + \frac{3G_*^2}{8} + \frac{6l^2}{\alpha^2} + 2l^2\,\mathrm{e}^{2kl}(G_*^2 + \varepsilon^2)\right]. \tag{2.4}$$

If we put

$$l = 1/k, \tag{2.5}$$

then

$$\|V(\Phi) - G\|_k^2 \leq \frac{4}{k}\,\mathrm{e}^2(G_*^2 + \varepsilon^2) \times \left[\frac{5}{2k} + \frac{3G_*^2}{8} + \frac{1}{k^2}\left(\frac{6}{\alpha^2} + 2\,\mathrm{e}^2(G_*^2 + \varepsilon^2)\right)\right]. \tag{2.6}$$

Therefore for every pair (G_*, ε) we choose $k_1 > 0$ such that for every $k \in (k_1, \infty)$

$$\Upsilon(k) = \frac{4}{k}\,\mathrm{e}^2(G_*^2 + \varepsilon^2)\left[\frac{5}{2k} + \frac{3G_*^2}{8} + \frac{1}{k^2}\left(\frac{6}{\alpha^2} + 2\,\mathrm{e}^2(G_*^2 + \varepsilon^2)\right)\right] \leq \varepsilon^2 \tag{2.7}$$

which means that $\Phi \in \Pi(G, \varepsilon, k)$ implies $V(\Phi) \in \Pi(G, \varepsilon, k)$.

Taking into account (1.22) and (1.23) we obtain

$$\|V(\Phi^{(1)}) - V(\Phi^{(2)})\|_k^2 \le \frac{\Phi_*^2}{k}\Big(\frac{16}{\alpha^2} + \frac{5l}{2} + 2\alpha^2 G_*^2 + \Phi_*^4\Big)\|\Phi^{(1)} - \Phi^{(2)}\|_k^2$$
$$\le \frac{2}{k}\,\mathrm{e}^2(G_*^2+\varepsilon^2)\|\Phi^{(1)} - \Phi^{(2)}\|^2\Big[\frac{16}{\alpha^2} + \frac{5}{2k} + 2\alpha^2 G_*^2 + 4\,\mathrm{e}^4(G_*^2+\varepsilon^2)\Big]. \quad (2.8)$$

It is clear that we can choose $k_2 \in (k_1, \infty)$ such that for every $k \in (k_2, \infty)$, we have $q(k) \in (0,1)$. Here

$$q(k) = \frac{2}{k}\,\mathrm{e}^2(G_*^2+\varepsilon^2)\Big[\frac{16}{\alpha^2} + \frac{5}{2k} + 2\alpha^2 G_*^2 + 4\,\mathrm{e}^4(G_*^2+\varepsilon^2)\Big].$$

Therefore we can put

$$l_* = \min(l_0, 1/k_2). \quad (2.9)$$

Then for every $l \in (0, l_*)$, the operator V has exactly one fixed point in $\Pi(G,\varepsilon,k)$. The stability can be proved by the same arguments as in the proof of Theorem 1.1. □

Appendix 1. PROOF OF LEMMA 1.1

Let $\Phi(x,t) \in L_2(l)$ and $\Phi_* = \|\Phi\|(l)$. Then we estimate every component of $V(\Phi)$:

$$|V(\Phi)_{(1)}| \le |G_{(1)}(x,t)| + \frac{1}{2}\Big[\int_0^x |\Phi_3(\xi)|^2 d\,\xi\Big]^{1/2}$$
$$\times \Big(\Big[\int_0^x |\Phi_1(\xi, t+x-\xi)|^2\,d\xi\Big]^{1/2} + \Big[\int_0^x |\Phi_1(\xi, t-x+\xi)|^2\,d\xi\Big]^{1/2}\Big),$$

$$V(\Phi)_{(1)}^2(x,t) \le 2|G_{(1)}(x,t)|^2$$
$$+ \int_0^x |\Phi_3(\xi)|^2\,d\xi \int_0^x \big[|\Phi_1(\xi, t+x-\xi)|^2 + |\Phi_1(\xi, t-x+\xi)|^2\big]\,d\xi, \quad (3.1)$$

$$\int_x^{2l-x} V(\Phi)_{(1)}^2(x,t)\,dt \le 2\int_x^{2l-x} |G_{(1)}(x,t)|^2\,dt$$
$$+ 2\|\Phi_3\|^2(x)\int_0^x \|\Phi_1\|_{(1)}^2(\xi)\,d\xi. \quad (3.2)$$

We recall that $\|\Phi_1\|^2_{(1)}(x) = \int_x^{2l-x} \Phi_1(x,t)\,dt$ and denote $G_* = \|G\|(l)$. Then we obtain from (3.2)

$$\|V(\Phi)_{(1)}\|^2_{L_2(l)} \le 2\|G_1\|^2_{L_2(l)} + 2\Phi_*^4, \tag{3.3}$$

$$|V(\Phi)_{(2)}|^2(x) \le \frac{1}{2}\int_0^x |\Phi_3(\xi)|^2\,d\xi \int_0^x |\Phi_2(\xi)|^2\,d\xi + 2|G_2(x)|^2, \tag{3.4}$$

$$\|V(\Phi)_{(2)}\|^2(l) \le 2\|G_2\|^2(l) + \Phi_*^4/2. \tag{3.5}$$

$$\begin{aligned}|V(\Phi)_{(3)}(x)| \le\ & |G_3(x)| \\ &+ \frac{1}{2}|\alpha|\,|G_3(x)|\Big[\int_0^x |\Phi_3(\xi)|^2\,d\xi\Big]^{1/2}\Big[\int_0^x |\Phi_2(\xi)|^2\,d\xi\Big]^{1/2} \\ &+ \frac{2}{|\alpha|}\Big[\int_0^x |\Phi_3(\xi)|^2\,d\xi\Big]^{1/2}\Big[\int_0^x |\Phi_1(\xi,2x-\xi)|^2\,d\xi\Big]^{1/2} \\ &+ \Big[\int_0^x |\Phi_3(\xi)|^2\,d\xi\Big]^{1/2}\Big[\int_0^x |\Phi_2(\xi)|^2\,d\xi\Big]^{1/2} \\ &\times \Big[\int_0^x |\Phi_3(\xi)|^2\,d\xi\Big]^{1/2}\Big[\int_0^x |\Phi_1(\xi,2x-\xi)|^2\,d\xi\Big]^{1/2}.\end{aligned} \tag{3.6}$$

$$\|V(\Phi)_{(3)}\|^2(l) \le 4\|G_3\|^2(l) + \alpha^2\Phi_*^4\|G_{(3)}\|^2(l) + 16\Phi_*^4/\alpha^2 + 4\Phi_*^8. \tag{3.7}$$

Appendix 2. PROOF OF LEMMA 1.2

Let $\Phi \in \Pi(G,\varepsilon,k))$. Let us prove that $V(\Phi) \in \Pi(G,\varepsilon,k))$

$$|V(\Phi)_{(1)} - G_1| \le |\mathcal{B}_1(\Phi)| \le \int_0^x |\Phi_3(\xi)||\Phi_1(\xi,t+x-\xi) + \Phi_1(\xi,t-x+\xi)|\,d\xi,$$

$$\begin{aligned}&\int_x^{2l-x} (V(\Phi)_{(1)} - G_1)^2(x,t)\,dt \\ &\le 2\|\Phi_3\|^2(x)\int_x^{2l-x}\int_0^x \big[|\Phi_1(\xi,t+x-\xi)|^2 + |\Phi_1(\xi,t-x+\xi)|^2\big]\,\mathrm{e}^{-2k\xi}\,\mathrm{e}^{2k\xi}\,d\xi \\ &\qquad \le 4\Phi_*^2\,\frac{\mathrm{e}^{2kx}-1}{2k}\int_0^l \|\Phi_1\|^2_{(1)}(\xi)\,d\xi = \frac{2\Phi_*^2}{k}\,\mathrm{e}^{2kx}\,\|\Phi_1\|_k^2,\end{aligned}$$

$$\|V(\Phi)_{(1)} - G_1\|_k^2 \le \frac{2\Phi_*^2}{k}\,l\|\Phi_1\|_k^2, \tag{3.8}$$

$$\Phi_* = \|\Phi\|(l), \quad G_* = \|G\|(l), \quad \Phi_k = \|\Phi\|_k(l), \qquad G_* = \|G\|_k(l),$$

$$|V(\Phi)_{(2)} - G_2| \le \frac{1}{2}\int_0^x |\Phi_3(\xi)\Phi_2(\xi)|\,d\xi \le \frac{1}{2}\Big[\int_0^x |\Phi_3(\xi)|^2\,d\xi\Big]^{1/2}\Big[\int_0^x |\Phi_2(\xi)|^2\,d\xi\Big]^{1/2},$$

$$\int_0^x (V(\Phi)_{(2)} - G_2)^2(\xi)\,\mathrm{e}^{-2k\xi}\,d\xi \le \frac{\Phi_*^2}{2k}\,\|\Phi_2\|_k^2,$$

$$\|V(\Phi)_{(2)} - G_2\|_k^2 \le \frac{\Phi_*^2}{2k}\,\|\Phi_2\|_k^2. \tag{3.9}$$

$$\begin{aligned}|V(\Phi)_{(3)} - G_3| &\le \frac{1}{2}|\alpha||G_3(x)|\Big[\int_0^x |\Phi_3(\xi)|^2\,d\xi\Big]^{1/2}\Big[\int_0^x |\Phi_2(\xi)|^2\,d\xi\Big]^{1/2}\\ &+ \frac{2}{|\alpha|}\Big[\int_0^x |\Phi_3(\xi)|^2\,d\xi\Big]^{1/2}\Big[\int_0^x |\Phi_1(\xi, 2x-\xi)|^2\,d\xi\Big]^{1/2}\\ &+ \int_0^x |\Phi_3(\xi)|^2\,d\xi\Big[\int_0^x |\Phi_2(\xi)|^2\,d\xi\Big]^{1/2}\Big[\int_0^x |\Phi_1(\xi, 2x-\xi)|^2\,d\xi\Big]^{1/2}\\ &\le \frac{1}{2}\,|\alpha||G_3(x)|\sqrt{\frac{\mathrm{e}^{2kx}}{2k}}\,\|\Phi_3\|_k\Phi_* + \frac{2}{|\alpha|}\sqrt{\frac{\mathrm{e}^{2kx}}{2k}}\,\|\Phi_3\|_k\Big[\int_0^x |\Phi_1(\xi, 2x-\xi)|^2\,d\xi\Big]^{1/2}\\ &+ \Phi_*^2\sqrt{\frac{\mathrm{e}^{2kx}}{2k}}\,\|\Phi_3\|_k\Big[\int_0^x |\Phi_1(\xi, 2x-\xi)|^2\,d\xi\Big]^{1/2},\end{aligned}$$

$$\|V(\Phi)_{(3)} - G_3\|_k^2 \le \frac{\Phi_*^2}{2k}\,\|\Phi_3\|_k^2\Big[\frac{3}{4}\,\alpha^2\|G_3\|^2(l) + \frac{12l}{\alpha^2} + \Phi_*^2 l\Big]. \tag{3.10}$$

By the definition of $\Phi_k = \|\Phi\|_k$ we obtain from (3.8)–3.10

$$\|V(\Phi) - G_k\|_k^2 \le \frac{l\Phi_*^2}{k}\Big[\frac{5}{2} + \frac{3}{8l}\,\|G\|^2(l) + \frac{6l}{\alpha^2} + l\Phi_*^2\Big]\Phi_k^2. \tag{3.11}$$

Therefore since $\Phi \in \Pi(G, \varepsilon, k)$, we have $V(\Phi) \in \Pi(G, \varepsilon, k)$ provided Φ_*, G_*, k satisfy the condition

$$\frac{l\Phi_*^2}{k}\Big[\frac{5}{2} + \frac{3}{8l}\,\|G\|^2(l) + \frac{6l}{\alpha^2} + l\Phi_*^2\Big](G_k + \varepsilon)^2 < \varepsilon. \tag{3.12}$$

Taking into account the property of k-norm

$$\|a\|_k^2(x) \le \|a\|^2(x) \le \mathrm{e}^{2kx}\,\|a\|_k^2(x) \tag{3.13}$$

we conclude that the condition

$$\frac{l}{k}\,(G_k + \varepsilon)\,\mathrm{e}^{2kl}\,\Big[\frac{5}{2} + \frac{3}{8l}\,G_k^2\,\mathrm{e}^{2kl} + \frac{6l}{\alpha^2}\,l(G_k + \varepsilon)^2\,\mathrm{e}^{2kl}\Big] < \varepsilon \tag{3.14}$$

will be justified if

$$\varepsilon^2 < \mathrm{e}^{-2kl}/2, \tag{3.15}$$

$$G_*^2 < \mathrm{e}^{-2kl}/2, \tag{3.16}$$

$$16l^2 + 2l + 3/2 < k. \tag{3.17}$$

Thus the proof of Lemma 1.2 is complete.

Acknowledgement

The first author appreciates a very kind support of all Faculty Members of Graduate School of Mathematics, University of Tokyo, and especially of Professor Masahiro Yamamoto during their investigations of inverse acoustic problem.

REFERENCES

1. S. I. Kabanikhin, K. T. Iskakov, and M. Yamamoto, H_1-conditional stability with explicit Lipschitz constant for a one-dimensional inverse acoustic problem. *J. Inv. Ill-Posed Problems* (2001) **9**, No. 3, 249–267.

2. V. G. Romanov, S. I. Kabanikhin, *Inverse Problems for Maxwell's Equations.* VSP, Utrecht, 1994.

3. J. S. Azamatov and S. I. Kabanikhin, Nonlinear operator equations. L_2-theory. *J. Inv. Ill-Posed Problems* (1999) **7**, No. 6, 497–529.

4. S. I. Kabanikhin and K. T. Iskakov, Justification of the steepest descent method for the integral statement of an inverse problem for a hyperbolic equation. *Sib. Math. J.* (2001) **42**, No. 3, 478–494.

5. V. G. Romanov and M. Yamamoto, Multidimensional inverse hyperbolic problem with impulse input and a single boundary measurement. *J. Inv. Ill-Posed Problems* (1999) **7**, No. 6, 573–588.

Ill-Posed and Non-Classical Problems of
Mathematical Physics and Analysis, Samarkand, 2000, pp. 71–80
M. M. Lavrent'ev and S. I. Kabanikhin (Eds)

Fourier series in Banach spaces

DJ. KHADJIEV* and A. ÇAVUŞ†

Abstract — Let T be the group $\{e^{it} \mid 0 \leq t < 2\pi\}$. We consider T with its euclidean topology. Our main results are as follows: 1) Theorem 1 which establishes a connection between spectrums of a continuous linear representation of T in a reflexive Banach space and its conjugate linear representation; 2) Theorem 11 which gives a description of all left (right) simple actions of T in Banach algebras; 3) Theorem 18 which establishes an analog of the theorem on integral for continuous linear representations of T in Banach spaces.

A part of results of the present paper was announced in an earlier our paper [1]. Further we use notions and notations of the book [2] and the paper [3].

Let H be a complex Banach space. We denote the group of all invertible bounded linear operators on H by $GL(H)$.

A homomorphism $\alpha : T \to GL(H)$ is called a linear representation of T in a Banach space H.

Let α be a linear representation of T in a Banach space H.

Definition 1. A point $x \in H$ is called a continuity point of α if $\lim_{t\to 0} \alpha(t)x = x$. A linear representation α is called continuous if $\lim_{t\to 0} \alpha(t)x = x$ for all $x \in H$.

Every continuous linear representation of T is equivalent to an isometric continuous linear representation [2, p. 83].

*National University of Uzbekistan, Tashkent, Uzbekistan
†Karadeniz Technical University, Trabzon, Turkey. E-mails: haciyev@risc01.ktu.edu.tr, cavus@jbsd.ktu.edu.tr

Let α be a continuous linear representation. Then there exists the Riemann's integral:

$$F_n(x) = \int_0^{2\pi} e^{-int}\alpha(t)x\,dt$$

for every $n \in Z$. It is easily seen that there exists a positive constant L such that $\|F_n(x)\| \leq L\|x\|$ for all $x \in H$ and $n \in Z$. We put $\mathrm{Spec}\,(x) = \{n \in Z \mid F_n(x) \neq 0\}$, $\mathrm{Spec}\,(H) = \bigcup_{x\in H} \mathrm{Spec}\,(x)$ and $H_n = \{x \in H \mid \alpha(t)x = e^{int}x, \forall t \in T\}$. H_n is a closed T-invariant linear subspace of H. Let H' be the conjugate linear space of H. Denote the conjugate linear operator of $\alpha(t)$ by $\alpha'(t)$. Then $\alpha' : T \to GL(H')$ is a linear representation of T in H'.

Proposition 1. *Let α be a continuous isometric linear representation of T in a Banach space H. Then:*

1) $F_n(\alpha(t)x) = \alpha(t)F_n(x) = e^{int}F_n(x)$ *for all* $t \in T, n \in Z$ *and* $x \in H$;

2) $F_n^2 = F_n$ *for all* $n \in Z$;

3) $\|F_n\| = 1$*for all* $n \in Z$;

4) $\mathrm{Im}\,F_n = \{y \in H \mid y = F_n(x),\ x \in H\} = H_n$ *for all* $n \in Z$;

5) $F_n \circ F_m = O$ *for all* $n \neq m,\ n, m \in Z$;

6) The subspaces $\{H_n \mid n \in Z\}$ *are linear independent.*

Let us put $s_n(x) := \sum_{k=-n}^{n} F_k(x)$ *and* $\sigma_n(x) := (s_0(x) + s_1(x) + \cdots + s_n(x))/(n+1)$.

Proposition 2. *Let α be a continuous isometric linear representation of T in a Banach space H. Then:*

1) $\sigma_n(x) = \dfrac{1}{2\pi}\displaystyle\int_T K_n(t)\alpha(t)x\,dt$ *for all* $x \in H$ *and* $n \in N$, *where the function* $K_n(t)$ *is Fejeŕs kernel and has the form* $K_n(t) = \sin^2((n+1)t/2)/[(n+1)\sin^2(t/2)]$

2) $\sup\|\sigma_n(x)\| \leq \|x\|$ *for all* $x \in H$.

Theorem 1. *Let α be a continuous linear representation of T in a Banach space H. Then* $\lim_{n\to\infty}\sigma_n(x) = x$ *for all* $x \in H$.

Theorem 1 is known for homogeneous Banach spaces of functions on T [4, p. 14].

Corollary 1. *Let α be a continuous isometric linear representation of T in a Banach space H.*

1) if $F_n(x) = 0$ for all $n \in Z$ then $x = 0$;

2) if $F_n(x) = F_n(y)$ for all $n \in Z$ then $x = y$.

This corollary is a generalization of the uniqueness theorem for Fourier series in $L^1(T)$ [4, p. 13].

Theorem 2. *Let α be a continuous linear representation of T in a Banach space H. Then $\lim_{n\to\infty} F_n(x) = 0$ for all $x \in H$.*

This theorem is a generalization of Riemann—Lebesque lemma in $L^1(T)$ [4, p. 13].

Theorem 3. *Let α be a continuous linear representation of T in a reflexive Banach space H. Then:*

1) α' is a continuous linear representation;

2) $\mathrm{Spec}\,(H) = \mathrm{Spec}\,(H')$;

3) $\dim_C H_n = \dim_C(H'_n)$ *for all $n \in Z$.*

Remark 1. Theorem 3 is not true if H is not reflexive. For example, let $H = L^1(T)$, $(\alpha(t)f)(x) = f(x+t)$. Then α' is not continuous.

Let $\hat{V} := \{(V_n, \|\cdot\|_n) \mid n \in Z\}$ be a family of complex Banach spaces. We shall consider sequences $x = \{x_n \mid n \in Z\}$, where $x_n \in V_n$.

A sequence $x = \{x_n \mid n \in Z\}$ is called bounded, if there exists $M > 0$ such that $\|x_n\|_n \leq M$ for all $n \in Z$. Denote the set of all bounded sequences by $m\{\hat{V}\}$. $m\{\hat{V}\}$ is a complex linear space with respect to the following operations:

$$\{x_n\} + \{y_n\} = \{x_n + y_n\},$$
$$\lambda\{x_n\} = \{\lambda x_n\}, \qquad \lambda \in C.$$

We define a norm in $m\{\hat{V}\}$ by

$$\|x\| = \sup_{n \in Z} \|x_n\|_n.$$

Then $m\{\hat{V}\}$ is a Banach space. We define a linear representation β of T in $m\{\hat{V}\}$ by $\beta(t)x = \{e^{int}x_n \mid n \in Z\}$.

Denote the set of all $x \in m\{\hat{V}\}$ such that $\lim_{|n|\to+\infty} \|x_n\|_n = 0$ by $m_0\{\hat{V}\}$.

Theorem 4. *$m_0\{\hat{V}\}$ is a closed T – invariant linear subspace of $m\{\hat{V}\}$, and $x \in m\{\hat{V}\}$ is a continuity point of the linear representation β if and only if $x \in m_0\{\hat{V}\}$.*

From the Proposition 1 we have $\sup_{n\in Z} \|F_n(x)\| \le \|x\|$ for all $x \in H$. Therefore, a description of all continuous isometric linear representations of T in a Banach space, satisfying the condition $\sup_{n\in Z} \|F_n(x)\| = \|x\|$ for all $x \in H$, is an interesting problem.

The following theorem is a solution of this problem.

Theorem 5. *Let α be a continuous isometric linear representation of T in a Banach space H and $H_n = \{x \in H \mid \alpha(t)x = e^{int}x,\ \forall t \in T\}$. Then:*

1) *The mapping $F : H \to m\{\hat{H}\}$, where $F(x) = \{F_n(x) \mid n \in Z\}$, is an injective bounded linear operator and $\overline{F(H)} = m_0\{\hat{H}\}$;*

2) *$F(H) = m_0\{\hat{H}\}$ iff $\sup_{n\in Z} \|F_n(x)\| = \|x\|$ for all $x \in H$.*

Let α be a continuous linear representaion in a Banach space H. We consider the following formal series:

$$\sum_{n=-\infty}^{+\infty} x_n, \tag{1}$$

where $x_n \in H_n$, $n \in Z$. For each integer number $n \ge 0$ let

$$s_n = \sum_{k=-n}^{n} x_k, \qquad \sigma_n = \frac{s_0 + s_1 + \cdots + s_n}{n+1}.$$

be the n-th partial sums and n-th Cesaro's means of the series (1).

Definition 2. We say that the series (1) is a Fourier series of the point $x \in H$ if $F_n(x) = x_n$ for all $n \in Z$.

Proposition 3. *Let $\alpha : T \to GL(H)$ be a continuous linear representation of T in a Banach space H. Then the series (1) is a Fourier series of an $x \in H$ if and only if the sequence $\{\sigma_n\}$is convergent in H.*

Theorem 6. *Let α be a continuous linear representation in a reflexive Banach space H. Then the series (1) is a Fourier series of a point $x \in H$ if and only if the sequence $\{\sigma_n\}$ of the series (1) is bounded.*

Remark 2. This theorem is known for L^p, $1 < p < \infty$, and symmetric functional spaces [7].

Let α be a continuous linear representation of T in a Banach space H. Let us denote the set of all sequences $\tilde{x} = \{x_n \mid n \in Z\}$, such that $x_n \in H_n = F_n(H)$, $n \in Z$, and $\sup_{n\in Z} \|\sigma_n(\tilde{x})\| < +\infty$, by $\tilde{H}_\sigma$, and define

$$\|\tilde{x}\| := \sup_{n\in Z} \|\sigma_n(\tilde{x})\| < +\infty. \tag{2}$$

Theorem 7. *Let α be a continuous linear representation of T in a reflexive Banach space H. Then*

i. $\tilde{H}_\sigma$ is a Banach space with respect to the norm (2);

ii. The mapping $F : H \to \tilde{H}_\sigma$, $F(x) := \{F_n(x) \mid n \in Z\}$, is an isometric isomorphism.

Let α be a continuous linear representation of T in a Banach space H. Let us denote the set of all sequences $\tilde{x} = \{x_n \mid n \in Z\}$, such that $x_n \in H_n = F_n(H)$, $n \in Z$, and $\sup_{n\in Z} \|s_n(\tilde{x})\| < +\infty$, by $\tilde{H}_s$, and define

$$\|\tilde{x}\| := \sup_{n\in Z} \|s_n(\tilde{x})\| < +\infty. \tag{3}$$

Proposition 4. *Let α be a continuous linear representation of T in Banach space H. Then H_s is a Banach space with respect to the norm (3).*

Definition 3. A point $x \in H$ is an s-point of α if $\lim_{n\to\infty} s_n(x) = x$, where $s_n(x) = \sum_{k=-n}^{n} F_k(x)$.

Theorem 8. *Let α be a continuous isometric linear representation of T in a reflexive Banach space H. If the sequence $\{s_n\}$ of the series (1) is bounded, then there exists an s-point $x \in H$ such that $F_n(x) = x_n$ for all $n \in Z$.*

Theorem 9. *Let $\alpha : T \to GL(H)$ be a continuous linear representation in a Banach space H, $x \in H$, $p(x,r) := \sum_{n=-\infty}^{+\infty} r^{|n|} F_n(x)$ and $0 \le r < 1$. Then*

1) $p(x,r) = \dfrac{1}{2\pi} \displaystyle\int_{-\pi}^{\pi} P_r(t)\alpha(t)x\,dt$, where $P_r(t) := (1 - 2r\cos t + r^2)/(1 - r^2)$ is the Poisson's kernel;

2) $\lim_{r\to 1} p(x,r) = x$.

Proposition 5. *Let $\alpha : T \to GL(H)$ be a continuous isometric linear representation of T in a Banach space H. Then $\|p(x,r)\| \le \|x\|$ for all $x \in H$.*

Proposition 6. *Let $\alpha : T \to GL(H)$ be a continuous linear representation of T in a Banach space H, and $\{y_n\}$ be a convergent sequence in H. If $\lim_{n\to\infty} y_n = y$ then $\operatorname{Spec}(y) \subset \bigcup_{n=1}^{\infty} \operatorname{Spec}(y_n)$.*

Proposition 7. *Let $\alpha : T \to GL(H)$ be a continuous linear representation of T in a Banach space H, $x \in H \setminus \{\theta\}$, and $[x]$ be the smallest closed $\alpha(T)$-invariant subspace of H containing x. Then $\operatorname{Spec}(x) = \operatorname{Spec}[x]$.*

Theorem 10. *Let $\alpha : T \to GL(H)$ be a continuous linear representation in a Banach space H, and $x \in H \setminus \{\theta\}$. Suppose that $\dim_C H_n \le 1$ for all $n \in Z$. Then $[x] = H$ if and only if $\operatorname{Spec}(x) = \operatorname{Spec}(H)$.*

Remark 3. Theorem 10 is an analog of the approximation theorem of Wiener [8, Ch. II].

Theorem 11. *Let $\alpha : T \to GL(H)$ be a continuous linear representation in a Banach space H. Suppose that $\dim_C H_n < +\infty$ for all $n \in Z$. Then a closed subset $A \subset H$ is compact if and only if:*

i) A is bounded, i.e., there exists a positive constant $K > 0$ such that $\|x\| \le K$ for all $x \in A$;

ii) for any $\varepsilon > 0$ there exists a natural number $N(\varepsilon)$ such that $\|\sigma_n(x) - x\| < \varepsilon$ for all $x \in A$ and $n \ge N(\varepsilon)$.

Theorem 12. *Let $\alpha : T \to GL(H)$ be a continuous linear representation in a Banach space H. Suppose that $\dim_C H_n < +\infty$ for all $n \in Z$. Then a closed subset $A \subset H$ is compact if and only if:*

i) A is bounded;

ii) for any $\varepsilon > 0$ there exists a positive $\delta(\varepsilon)$ such that $\|\alpha(t)(x) - x\| < \varepsilon$ for all $x \in A$ and $|t| < \delta(\varepsilon)$.

Remark 4. For the space $H = L^p([0, 2\pi])$, $1 \le p < +\infty$, the sufficiency of the Theorem 12 due to Riesz [9, p. 69].

Theorem 13. *Let $\alpha : T \to GL(H)$ be a continuous linear representation of T in a Banach space H. Suppose that $\dim_C H_n < +\infty$ for all $n \in Z$. Then a closed subset $A \subset H$ is compact if and only if for any $\varepsilon > 0$ there exists a natural number $N(\varepsilon)$ such that $\|\frac{n}{n+1}\sigma_n(x) - x\| < \varepsilon$ for all $x \in A$ and $n \ge N(\varepsilon)$.*

Theorem 14. *Let $\alpha : T \to GL(H)$ be a continuous unitary linear representation of T in a Hilbert space H. Then:*

i) $\sum_{n=-\infty}^{+\infty} \|F_n(x)\|^2 = \|x\|^2$ *for all* $x \in H$;

ii) $\|s_n\| = 1$ *for all* $n \in N$;

iii) $\lim_{n \to +\infty} s_n(x) = x$ *for all* $x \in H$.

Definition 4. Let $\alpha : T \to GL(H)$ be a continuous linear representation of T in a Banach space H ,V be a linear subspace of H, and $x \in V$. x is said to be $\alpha(T)$-invariant if $\alpha(t)x = x$ for all $t \in T$.

Let us denote the set of all $\alpha(T)$-invariant elements of V by $V^{\top}$.

Definition 5. Let $\alpha : T \to GL(H)$ be a continuous linear representation of T in a Banach space H and V be a linear subspace of H. V is said to be $\alpha(T)$-invariant if $\alpha(T)V \subset V$ for all $t \in T$.

Proposition 8. *Let $\alpha : T \to GL(H)$ be a continuous linear representation of T in a Banach space H and $\{V_\tau \mid \tau \in S\}$ be a family of $\alpha(T)$-invariant subspaces of H. Then* $\overline{\left(\sum_{\tau \in S} V_\tau^{\top}\right)} = \overline{\left(\sum_{\tau \in S} V_\tau\right)}^{\top} = \left(\overline{\sum_{\tau \in S} V_\tau}\right)^{\top}$.

This proposition is a topological analog of the statement (d) of the theorem in [5, Ch. II, 3.2].

Let H be a Banach algebra and $Aut(H)$ be the group of all bounded automorphisms of H.

Definition 6. A homomorphism $\alpha : T \to Aut(H)$ is called an action of T in a Banach algebra H.Let $H^{\top} = \{x \in H \mid \alpha(t)x = x, \forall t \in T\}$ be the set of all $T-$ *invariant* elements of H. It is well-known that $H^{\top}$ is a closed subalgebra of H. Denote the complete lattice of all closed left (right) ideals of H by $L_l(H)(L_r(H))$ and the complete lattice of all T-invariant closed left (right) ideals of H by $L_l^{\top}(H)(L_r^{\top}(H))$, respectively. Denote the complete lattice of all closed left (right) ideals of the subalgebra $H^{\top}$ by $L_l(H^{\top})(L_r(H^{\top}))$. For $I \in L_l(H^{\top})$ put $I^e := \overline{HI}$. I^e is the smallest closed left ideal of H containing I and T-invariant. For $A \in L_l(H)$ put $A^C := A \cap H$ and $I^{ec} := (I^e)^C$.

Proposition 9. *Let α be a continuous action of T in a Banach algebra H, and $F_0 : H \to H$ be the linear operator defined by* $F_0(x) := \dfrac{1}{2\pi}\displaystyle\int_T \alpha(t)x\,dt$. *Then* $F_0(ax) = aF_0(x)$ *and* $F_0(xa) = F_0(x)a$ *for all* $a \in H^{\top}, x \in H$.

This proposition shows that F_0 is an analog of Reynold's operator in invariant theory [6, p. 27].

Theorem 15. *Let $\alpha : T \to Aut(H)$ be a continuous action of T in a Banach algebra H. Then*

i) $I^{eC} = I$ for all $I \in L_l(H^\top)$;

ii) the mapping $e : L_l(H^\top) \underset{I \to I^e}{\to} L_l^\top(H)$ is injective.

Remark 5. An algebraic version of Theorem 15 was obtained for actions of reductive groups in polynomial rings [10, 11].

Let $M(H)$ be the set of all maximal ideals of a commutative Banach algebra H.

Proposition 10. *Let α be a continuous action of T in a commutative Banach algebra H. Then:*

i) if $I \in M(H)$ then $I^C \in M(H^\top)$;

ii) the mapping $c : M(H) \underset{I \to I^C}{\to} M(H^\top)$ is surjective.

Definition 7. An action $\alpha : T \to Aut(H)$ of T in a Banach algebra H is called left (right) simple if $L_l^\top(H) = \{0, H\}$ ($L_r^\top(H) = \{0, H\}$).

Theorem 16. *Let $\alpha : T \to Aut(H)$ be a continuous action of T in a Banach algebra H. Then α is left simple if and only if:*

i) Spec (H) *is a subgroup of Z;*

ii) $\dim_C H_n \leq 1$ for all $n \in Z$;

iii) $H_n H_m = H_{n+m}$ for all $m, n \in$ Spec (H).

Definition 8. Let $\alpha : T \to GL(H)$ be a continuous linear representation in a Banach space H, $x \in H$, and $t_0 \in T$. If the limit $d_{t_0}^\alpha x := \lim_{t \to 0}(\alpha(t)x - \alpha(t_0)x)/(t - t_0)$ exits then x is said to be differentiable at $t_0 \in T$ with respect to α, and $d_{t_0}^\alpha x$ is called the derivative of x at $t_0 \in T$ with respect to α.

Proposition 11. *Let $\alpha : T \to GL(H)$ be a continuous linear representation in a Banach space H, $x \in H$, and $t_0 \in T$. Then x is differentiable at $t_0 \in T$ with respect to α if and only if x is differentiable at $0 \in T$ with respect to α. Furthermore $d_{t_0}^\alpha x = \alpha(t_0) d_0^\alpha x$.*

Definition 9. Let $D(\alpha)$ be the set of all points of H,differentiable at $0 \in T$ with respect to α.

Theorem 17. *Let $\alpha : T \to GL(H)$ be a continuous linear representation in a Banach space H. Then*

1) $D(\alpha)$ is dense in H;

2) $D(\alpha) = H$ if and only if $\mathrm{Spec}\,(H)$ is a finite set;

3) $D(\alpha)$ is a linear subspace of H;

4) $F_n(d_0^\alpha x) = \mathrm{i}n F_n(x)$ for all $n \in Z$ and $x \in D(\alpha)$.

Theorem 18. *Let $\alpha : T \to GL(H)$ be a continuous linear representation in a Banach space H and $a \in H$. Then the differential equation $d_0^\alpha x = a$ has a solution if and only if $F_0(a) = 0$.*

Remark 6. This theorem is an analog of the theorem about integral [12,ch.VI].

REFERENCES

1. Dj. Khadjiev and A. Çavuş, The imbedding theorem for continuous linear representations of the rotation group of the circle in Banach spaces. *Dokl. Acad. Nauk Resp. Uzbekistan* (2000) **7**, 8–11.

2. Y. I. Lyubich, *Introduction to the Theory of Banach Representations of Groups.* Birkhauser Verlag, Berlin, 1988.

3. Dj. Khadjiev and R. G. Aripov, Linear representations of the rotation group of the circle in local convex spaces. *Dokl. Acad. Nauk Resp. Uzbekistan* (1997) **9**, 8–11.

4. Y. Katznelson, *An Introduction to Harmonic Analysis.* Dover Publ., New York, 1976.

5. T. H. Kraft, *Geometrische Methoden in der Invarianten-Theorie.* Friedr. Viewag & Sohn Braunschweig, Wiesbaden, 1985.

6. D. Mumford, J. Fogarty, and F. Kirwan, *Geometric Invariant Theory.* Springer-Verlag, Berlin—New York, 1994.

7. Dj. Khadjiev and U. Karakbaev, The description of types of Fourier series in symmetric spaces. *Dokl. Acad. Nauk Resp. Uzbekistan* (1991) **7**, 11–13.

8. N. Wiener, *The Fourier Integral and Certain of its Applications.* Dover Publ., New York, 1933.

9. L. A. Lusternik and V. I. Sobolev, *Elements of Functional Analysis.* Gordan and Breach, Science Publ., New York, 1968.

10. M. Nagata, Invariants of a group in an affine ring. *J. Math. Kyoto Univ.* (1964) **3**, No. 3, 369–377.

11. Dj. Khadjiev, About properties of extension and contraction operations of ideals of fixed algebras of reductive groups in a positive characteristic. *Funct. Anal. Appl.* (1991) **18**, No. 4, 273–280.

12. B. M. Levitan and V. V. Jikov, *Almost Periodic Functions and Differential Equations.* Moscow State Univ. Publ., Moscow, 1978.

Ill-Posed and Non-Classical Problems of
Mathematical Physics and Analysis, Samarkand, 2000, pp. 81–101
M. M. Lavrent'ev and S. I. Kabanikhin (Eds)

Ill-posed and inverse problems for hyperbolic equations

M. M. LAVRENT'EV*

1. The notions of ill-posedness for boundary-value problems for equations of mathematical physics and ill-posed problems were introduced in the beginning of 20th century by the French mathematician J. Hadamard [23, 24]. Further, these notions were revised and set in most textbooks of mathematical physics [5, 13, 14].

At first it had seemed that ill-posed problems do not describe any real physical phenomena. This point of view was universally adopted for years.

The first who begin to consider the concept of ill-posed problems was A. N. Tikhonov [16]. Starting in 60th, the theory of ill-posed problems becomes of an interest of scientists. It had turned out that this theory is connected with a lot of various applied problems.

A considerable contribution to the theory of ill-posed problems was done by A. N. Tikhonov and his disciples (Moscow school), M. M. Lavrent'ev and his disciples (Novosibirsk school), V. K. Ivanov and his followers (Ekaterinburg school) [3, 6, 7, 17, 18]. The groups of scientists who develop the theory of ill-posed problems are in republics of Soviet Union and in other countries [4, 9, 15, 21, 22, 26, 27].

In this report we shall tell about some directions in the ill-posed problem theory which are poorly known.

The classical ill-posed boundary-value problems for differential equations are the Cauchy problems. First of all, it is the Cauchy problem for the Laplace equation and the Cauchy problem for the heat conductivity equation with inverse time. The Cauchy problems for hyperbolic and parabolic

*Sobolev Institute of Mathematics, Siberian Branch of Russian Academy of Science, Novosibirsk

equations with the data given in time-like manifolds are ill-posed problems also. All these problems are connected with the problems of interpreting the data of physical measurements.

2. We consider the Dirichlet problem for the simplest hyperbolic equation — the d'Alembert equation. We need find the solution $u(x,t)$ of the equation

$$\frac{\partial^2 u}{\partial t^2} = \frac{\partial^2 u}{\partial x^2} \tag{1}$$

in the rectangle

$$0 < x < 1, \qquad 0 < t < l,$$

satisfying the boundary conditions

$$\begin{gathered} u(0,t) = u(1,t) = 0, \\ u(x,0) = 0, \qquad u(x,l) = f(x). \end{gathered} \tag{2}$$

In order that the solution of this problem be unique it is necessary and sufficient that the number l be irrational.

3. The Dirichlet problem for the d'Alembert equation was considered by several authors. Certain results on this problem can be found in [25]. In our paper the d'Alembert equation is considered in the form

$$\frac{\partial^2 u}{\partial x\, \partial y} = 0 \tag{3}$$

for the domains convex in the directions of the x-axis and y-axis. This means that the boundary of such domains is a Gordan curve which has no more than two common points with each curve parallel to x–axis or y–axis.

The author of [25] connects the research of the Dirichlet problem with the mappings of the domains of the (x, y)–plane onto certain canonical domains of the (u, v)–plane of the form

$$u = \varphi(x), \qquad v = \psi(y), \tag{4}$$

where $\varphi(x)$ and $\psi(y)$ are continuous monotone functions.

If the functions $\varphi(x)$, $\psi(y)$ are continuously differentiable, the solutions of the d'Alembert equation are invariant with respect to these mappings.

The analogous situation holds true in the case of the Dirichlet problem for the Laplace equation and for conformal mappings.

In this paper no applied problems connected with this problem are mentioned.

4. We observe the two scientific directions connected with mathematical description of physical phenomena, where the Dirichlet problem for the d'Alembert equation and the mappings of form (4) are used.

The first direction is the research of boundary-value problems for the system of equations which describes the small vibrations of rotating liquid. The first settings and results on this problem belong to S. L. Sobolev. However, the first steps here were connected with the defense theme; as a result, the first publication on this subject belongs to Sobolev's disciple R. A. Aleksandryan [1]. Further, a lot of works of S. L. Sobolev and his followers had appeared. The most of these results can be found in the monograph of S. L. Sobolev and T. I. Zelenyak [2].

The second direction is connected with gas dynamics. First settings belong to M. A. Lavrent'ev. The first publications in this direction belong to B. V. Shabat and M. M. Lavrent'ev [8, 19]. Further development was done by M. A. Lavrent'ev and B. V. Shabat [20, 10].

It is known, that the two–dimensional stationary flow of ideal gas is described by the system of elliptic equations; the supersonic flow is described by the hyperbolic system.

The simplest system of elliptic equations is the Cauchy—Riemann system

$$\frac{\partial u}{\partial x} = \frac{\partial v}{\partial y}; \quad \frac{\partial u}{\partial y} = -\frac{\partial v}{\partial x}. \tag{5}$$

The solutions of the Cauchy—Riemann system realize the conformal mappings.

The simplest system of hyperbolic equations is the d'Alembert system

$$\frac{\partial u}{\partial y} = 0, \quad \frac{\partial v}{\partial x} = 0.$$

The solutions of this system are the mappings of form (4). In [10], the mappings of this form are called the h-conformal mappings. These mappings we shall call by h-mappings.

5. We shall cite now the results of the work [10] on existence of h-mappings of the two classes of domains onto the canonical domains.

The mappings of the half-plane type. Let $f(x)$ be a continuously differentiable function such that

$$0 < a < f'(x) < b, \quad f(0) = 0,$$

where a, b are certain constants.

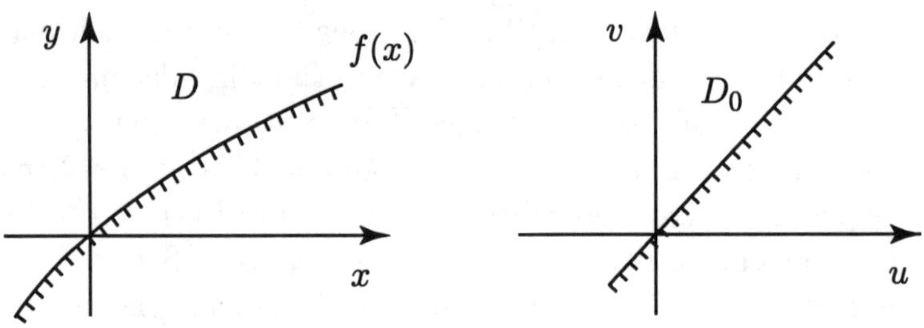

Figure 1

We consider the following domain in the plane (x, y):

$$D = \{(x, y) \mid y > f(x)\}.$$

Theorem 1. *There exists an h-mapping of the domain D onto the half-plane*

$$D_0 = \{(u, v) \mid v > u\}$$

satisfying the condition

$$u(0,0) = v(0,0) = 0,$$

i.e., leaving the origin immovable (Figure 1).

The mappings of the strip type. Let $f_1(x)$, $f_2(x)$ be continuously differentiable functions satisfying the conditions

$$0 < a < f_k'(x) < b, \quad k = 1, 2, \qquad f_2(x) > f_1(x) + c,$$

where a, b, c are certain constants, $c > 0$.

We consider the domain

$$D = \{(x, y) \mid f_1(x) < y < f_2(x)\}.$$

Theorem 2. *There exists an h-mapping of the domain D on the strip*

$$D_0 = \{(u, v) \mid u < v < u + 1\}$$

satisfying the condition

$$u(0,0) = v(0,0) = 0$$

(see Figure 2).

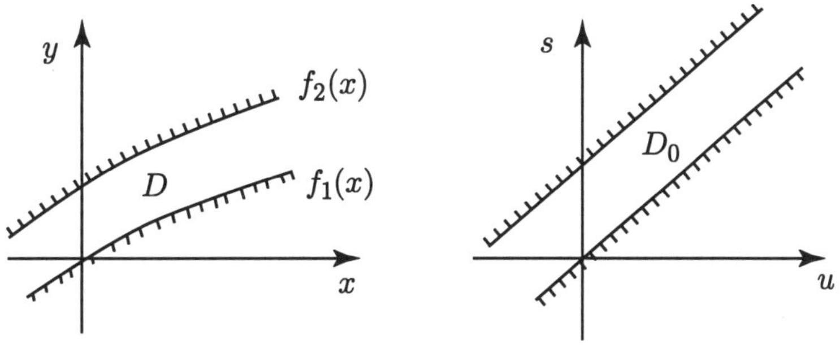

Figure 2

6. Mappings of bounded domains.
We shall give here some definitions and the results from [25] relatively h-mappings of bounded domains.

Let C be a Gordan curve convex relatively the x- and y-axes. The point P for which the lines parallel to the x-axis or to y-axis and coming through this point have no more points common with this curve is called the vertex of the curve C. The curve C may have one, two, three or four vertices.

Let P be a point in the curve C with the coordinates (x, y):

$$P(x, y) \in C.$$

We denote by AP the point on the curve C with the coordinates (x_1, y) and by T the transformation of the curve C on itself:

$$TP = BAP.$$

The vertices C are immovable points of the mappings A or B. The sequence of points P, AP, TP, ..., T^kP and the segments connecting these points we call λ-polygon defined by the point P. If there exists such n that $T^nP = P$, then the minimal n we call by the period of the point P. The point P in this case is called the P-periodic point of the curve C. The mapping T we shall call even if it conserves the positive direction on C.

For the curves C, relatively the transposition T, the four cases are possible.

1. All the points of C are periodic. The curve C, in this case, we shall call periodic.

2. The curve C contains both periodic and nonperiodic points. Such curve we shall call half-periodic.

3. There are no periodic points in the curve C. There exist no points P such that the set P, TP, ..., T^kP, ... is everywhere dense in C. We call the curve C in this case intransitive.

4. There are no periodic points in the curve C, but there exists a point P, so that the set P, TP, ..., T^kP, ... is everywhere dense in C. In this case, we shall call the curve C-transitive.

In the case 1 all the points have the common period n. If P is an arbitrary point, then the points $P, TP, \dots, T^{n-1}P$ divide C on n nonoverlapping arcs.

In the case 2 all periodic points have the common period n. The set F of all periodic points is closed. The additional set $C \setminus F$ is a countable set of arcs with the ends belonging to F. Each of these arcs is invariable under the action of the operator T^n.

We consider now the case 3. Let Q be an arbitrary point from C and σ be the set of limit points of the set $Q, TQ, \dots, T^kQ, \dots$. The set σ is perfect, nowhere dense, and is independent on the point Q.

In the case 4 the transformation T is equivalent topologically to rotation of a circle. This means that there exists a real ξ and a continuous mapping

$$t = f(P)$$

of the points P of the curve C on the points $e^{2\pi it}$ of a unit circle in the complex plane so that

$$f(TP) = t + \xi.$$

The constant ξ is irrational and is defined uniquely by the curve C. This constant is called by the modulus of the curve C.

Theorem 3. *Let C_1, C_2 be transitive curves with the common modulus ξ; D_1, D_2 be the domains bounded by these curves. Then there exists the h-mapping of the domain D_1 onto the domain D_2.*

Corollary 1. *Let C be a transitive curve with the modulus ξ and D be the domain bounded by C. Then there exists the h-mapping of the domain D onto the rectangle with the sides*

$$\begin{aligned} &y = -x + 1 && x \in [0, 1], \\ &y = x - 1 && x \in [1, \eta], \\ &y = x + 1 && x \in [0, \eta - 1], \\ &y = -x + 2n - 1 && x \in [\eta^{-1}, \eta], \qquad \eta = 1/(1 - \xi) \end{aligned}$$

(see Figure 3).

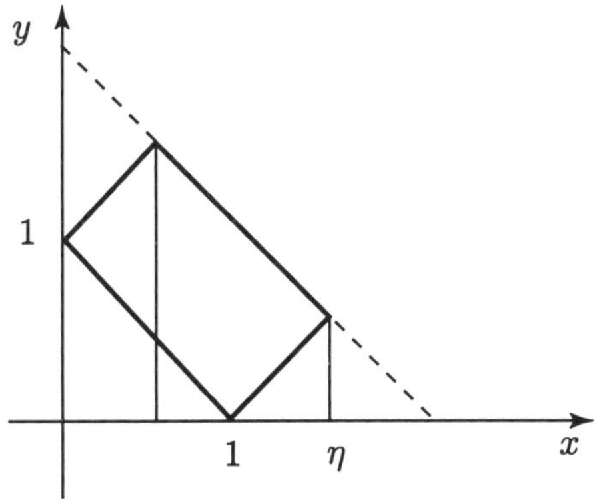

Figure 3

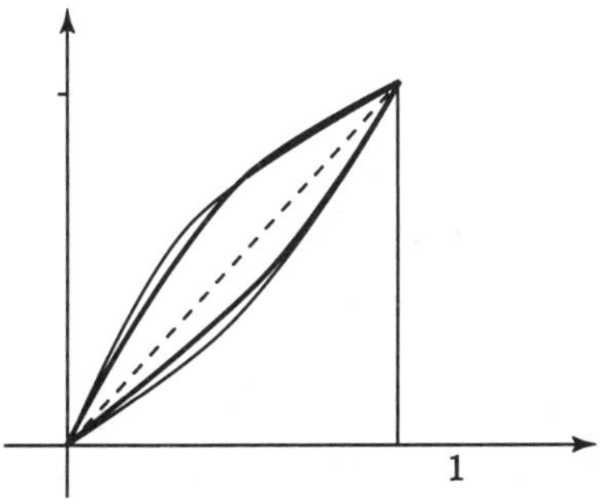

Figure 4

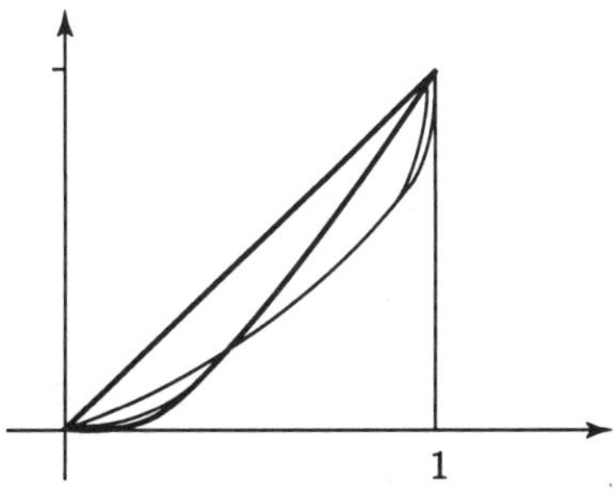

Figure 5

7. Mappings of domains with two vertices.
It is easy to see that the domain which has two vertices can be transformed to the domain

$$D = \{(x, y) \mid f_1(x) < y < f_2(x), \;\; x \in (0, 1)\}$$

with the aid of the h-mapping. The functions $f_k(x)$, $k = 1, 2$ satisfy the conditions:

1) $f_k(0) = 0$, $f_k(1) = 1$;

2) the functions $f_k(x)$ are continuous and monotonically increase;

3) $f_1(x) < f_2(x)$, $0 < x < 1$.

The domains of such type we shall call by the domains of the gap type (see Figure 4).

Theorem 4. *For each domain D of the gap type, there exists the h–mapping of this domain onto the domain*

$$D_0 = \{(x, y) \mid x^2 < y < x, \;\; x \in (0, 1)\}$$

(see Figure 5).

Proof. Let h be the h-mapping realized by the functions

$$x_1 = \varphi_1(x) = f_2(x), \qquad y_1 = y.$$

This mapping transfers the domain D into the domain

$$D_1 = \{(x_1, y_1) \mid f_3(x_1) < y_1 < x_1\},$$
$$f_3(x_1) = f_1[\tilde{f}_2(x_1)], \qquad \tilde{f}_2[f_2(x)] = x.$$

Consider now the h-mapping of the domain D_1 realized by the functions

$$x_2 = \varphi_2(x_1), \qquad y_2 = \varphi_2(y_1),$$

where the function $\varphi_2(x_1)$ is defined from the relations:

$$\varphi_2\{f_3[\tilde{\varphi}_2(x_2)]\} = x_2^2, \qquad \tilde{\varphi}_2[\varphi_2(x_1)] = x_1.$$

Evidently, the h-mapping consisted from these two h–mappings applied consequently transfers the initial domain D into the domain D_0 defined in the theorem formulation. □

The similar result may be obtained for the domains which have three vertices.

8. Mappings of the regular domains.

Let D be a domain convex relatively the x-axis and the y-axis, which have four vertices.

Denote the coordinates of these vertices by (a_1, b_1), (a_2, b_2), (a_3, b_3), (a_4, b_4). Let

$$a_1 < a_2 \leq a_3 < a_4, \qquad b_2 < b_1 \leq b_4 < b_3.$$

We call the domain regular if

$$a_2 = a_3, \qquad b_1 = b_4.$$

It is easy to see in this case that there exists the h-mapping of the domain D onto the domain, where

$$a_1 = -1; \quad a_2 = a_3 = 0; \quad a_4 = 1,$$
$$b_2 = -1; \quad b_1 = b_4 = 0; \quad b_3 = 1.$$

(see Figure 7).

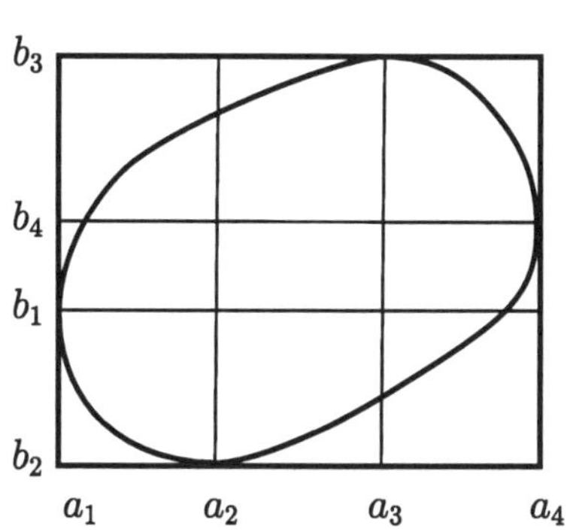

Figure 6

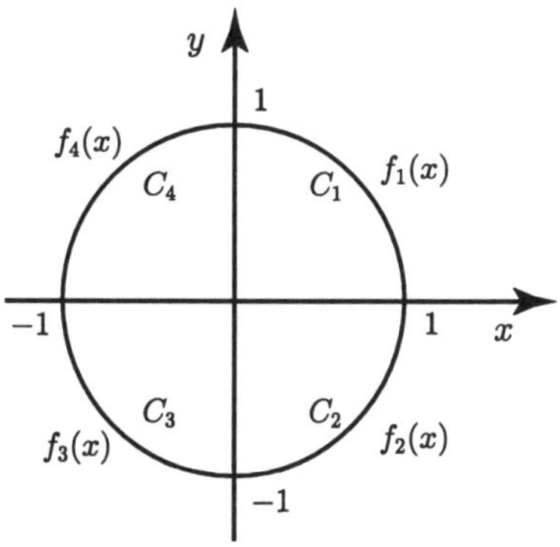

Figure 7

So we have supposed that D is the domain bounded by the curve C consisted from the four parts $C = C_1 \cup C_2 \cup C_3 \cup C_4$:

$$
\begin{aligned}
C_1 &= \{(x,y) \mid y = f_1(x),\ x \in [0,1],\ f_1(0) = 1;\ f_1(1) = 0\};\\
C_2 &= \{(x,y) \mid y = f_2(x),\ x \in [0,1],\ f_2(0) = -1;\ f_2(1) = 0\};\\
C_3 &= \{(x,y) \mid y = f_3(x),\ x \in [-1,0],\ f_3(-1) = 0;\ f_3(0) = -1\};\\
C_4 &= \{(x,y) \mid y = f_4(x),\ x \in [-1,0],\ f_4(-1) = 0;\ f_4(0) = 1\}.
\end{aligned}
$$

All the functions $f_k(x)$ are monotone and continuous.

We consider the h-mapping of the domain D so that

$$x_1 = \varphi^1(x), \qquad y_1 = \psi^1(y),$$

where the functions φ^1 and ψ^1 are defined as follows:

1°. $\varphi^1(x) = \varphi_1^1(x) = x$, $x \in [0,1]$; $\psi^1(y) = \psi_1^1(y)$, $y \in [0,1]$; $\psi_1^1[f_1(x)] = -x + 1$;

2°. $\psi^1(y) = \psi_2^1(y)$, $y \in [-1,0]$; $\psi_2^1[f_2(x)] = x - 1$, $x \in [0,1]$;

3°. $\varphi^1(x) = \varphi_3^1(x)$, $x \in [-1,0]$; $\psi_2^1(y) = f_3(x)$, $y = \tilde{f}_3(x)$; $\psi_2^1[\tilde{f}_3(x)] = f_3(x)$; $-\varphi_3^1(x) - 1 = \tilde{f}_3(x)$;

4°. $\psi^1(y) = \psi_4^1(y) = \psi_1^1(y)$, $y \in [0,1]$; $\psi_4^1(y) = f_4[\psi_4^1(x)]$, $y = f(x)$; $\psi_4^1[f(x)] = f_4[\psi_4^1(x)]$.

This mapping transfers the domain D into the domain D_0 of the plane (x_1, y_1) bounded by the three sides of the square and the curve $y_1 = f(x_1)$ (see Figure 8).

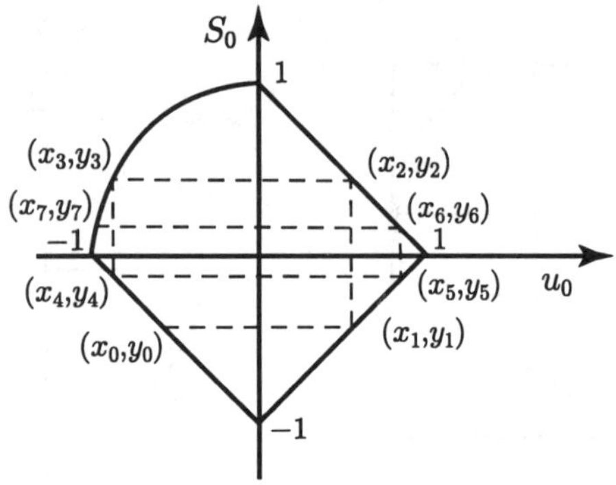

Figure 8

Consider now the h-mapping transferring the domain D_0 into the domain D_1 of the plane (u, v) of the type similar to D_0. We mean the domains bounded by three sides the square and by the certain curve

$$v = F(u), \qquad u \in [-1, 0],$$
$$F(-1) = 0, \quad F(0) = 1, \quad F'(u) > 0 \qquad u \in (-1, 0).$$

Let this mapping be defined by the functions

$$u = \varphi(x_1), \qquad v = \psi(y_1).$$

Since the mapping transfers D_0 into D_1, the functions are connected by the relations

$$\psi(1 - x) = -\varphi(x) + 1, \qquad x \in [0, 1],$$
$$\psi(x - 1) = \varphi(x) - 1, \quad \varphi(x) = -\psi(-x - 1) - 1, \quad x \in [-1, 0].$$

Thus, the values of the functions $\varphi(x)$, $\psi(x)$ in the segment $[-1, 1]$ are defined by the values of the function $\varphi(x)$ in the segment $[0, 1]$.

We suppose now that the functions $f(x)$ and $F(u)$ which define the parts of boundaries of the domains D_0 and D_1 satisfy the inequalities

$$f(x) > x - 1, \quad x \in (-1, 0), \qquad F(u) > u - 1, \quad u \in (-1, 0).$$

The functions $f(x)$, $F(u)$, $\varphi(x)$ are connected by the functional relation

$$f[\varphi(1 - x) - 1] = F[\varphi(1 - x) - 1].$$

Now, we consider that the functions $f(x)$, $F(u)$ are given and this relation is a certain functional equation relatively the function $\varphi(x)$. It is easy to show that if the conditions 1° and 2° hold, this equation has a unique solution satisfying the conditions

$$\varphi(0) = 1, \quad \varphi(1) = 0, \quad \varphi'(x) < 0, \quad x \in (0,1).$$

The function $F(u)$, for example, may be taken as follows:

$$F(u) = \sqrt{1+u}; \qquad u \in [-1,0].$$

Thus, we have established the following theorem.

Theorem 5. *For each regular domain D of the plane (x, y), for which the function $f(x)$ satisfies the inequality* (5), *there exists the h-mapping of the domain D onto the domain D_1 of the plane (u, v) bounded by the three sides of the square and the part of the parabola*

$$v = \sqrt{1+u}.$$

Thus, we have established the theorems on existence of h-mappings onto the canonical domains for the two classes of unbounded domains and three classes of bounded domains convex with respect to the x and y-axes. It seems to be interesting to research the possibility of h-mapping of an arbitrary domain on the canonical domains.

9. In Section 6 we had classified the Gordan curves bounded convex domains given in paper [25]. It is easy to see that the curves bounded the domains of gap-type are half-periodic; the periodic points with the period 1 are the two vertices of the curve. The curves restricted the regular domains may be either periodic with the period equal to 2 or half-periodic. In the first case there exists an h-mapping of the domain onto the square. In the second case, the period for periodic points is equal to 2; the vertices of the domain are periodic points. If the domain satisfies the conditions of Theorem 5, the periodic points are only the vertices.

10. We pass now to the Dirichlet problem for bounded convex domains. We restrict ourselves by the problems for the classes of domains considered in Theorems 3, 4.

So, let, at first, D be a domain defined in Theorem 3. Let $g_1(x)$, $g_2(x)$ be continuous functions defined in the segment $[0,1]$ and such that

$$g_1(0) = g_2(0), \qquad g_1(1) = g_2(1).$$

We need to find the solution of the d'Alembert equation

$$\frac{\partial^2 u}{\partial x\, \partial y} = 0$$

satisfying the conditions

$$u(x, f_1(x)) = g_1(x), \qquad u(x, f_2(x)) = g_2(x). \tag{6}$$

We shall consider the generalized solutions of the equation; i.e., each function

$$u(x, y) = \varphi(x) + \psi(y),$$

where $\varphi(x)$ and $\psi(y)$ are continuous, we shall consider as the solution of this equation.

Theorem 6. *Solution of problem* (3), (6) *exists and is unique.*

Proof. Solution of the problem formulated above is equivalent to solution of the following system of functional equations relatively the functions φ and ψ:

$$\varphi(x) + \psi[f_1(x)] = g_1(x), \qquad \varphi(x) + \psi[f_2(x)] = g_2(x). \tag{7}$$

From this system we obtain

$$\psi[f_2(x)] - \psi[f_1(x)] = g_2(x) - g_1(x). \tag{8}$$

From the conditions imposed on the functions $f_1(x)$, $f_2(x)$ it follows that the solution of functional equation (8) relatively the function ψ exists and is unique. Hence the statement of theorem follows. Evidently, in the space C of continuous functions, the solutions of system (7) depend on the functions $g_1(x)$, $g_2(x)$ continuously. Thus, the Dirichlet problem is well-posed in classical sense for the d'Alembert equation in the domains of gap type. □

We consider now the Dirichlet problem for the domain bounded by transitive curve. By Theorem 3, this problem is equivalent to solution of the d'Alembert equation in form (1) satisfying boundary-value conditions (2) for $l = \xi/(1 - \xi)$, where ξ is the modulus of the curve.

It is known that the solution of equation (1) with boundary conditions (2) may be represented in the form

$$u(x, t) = \sum_{k=1}^{\infty} a_k \sin k\pi x \cdot \sin k\pi t,$$

where

$$a_k = \frac{2}{\sin k\pi l} \int_0^1 f(x) \sin k\pi x \, dx.$$

Since the number l is irrational, we have

$$\sin k\pi l \neq 0.$$

We shall give now one of the notions of ill-posed problem theory.

If, for the boundary-value problem, there exists a pair of functional spaces in the definition of the norms, where a finite number of derivatives is used and the problem is well-posed in this pair of spaces, then the problem is called weakly ill-posed. If there are no such spaces, then the problem is strongly ill-posed.

The Cauchy problem for the Laplace equation and the other Cauchy problems mentioned in the beginning of this paper are strongly ill-posed. The classical weakly ill-posed problem is the problem of differentiation.

Since the number $\sin k\pi l$ may attain the extremely small values, the Dirichlet problem will be ill-posed in the pairs $L_2 \to L_2$ and $C \to C$. The type of instability of this problem depends on the properties of the number l. As it was noted in [25], if the number l is the Liouville number, i.e., for each n, there exist such integers p and q that

$$|l - p/q| < 1/q^n,$$

the problem will be strongly ill-posed. If there exist such numbers $c > 0$, $\sigma > 0$ that for each k the inequality

$$|\sin k\pi l| > Ck^{-\sigma}$$

holds, then the problem will be weakly ill-posed and the inequality

$$\|u\|_{L_2} \leq C\|f\|_{W_2^m}, \qquad m > \sigma$$

holds. The analogous condition is cited in [7] relatively the following problem for the Laplace equation:

find the function $u(x, y)$ which solves the Laplace equation

$$\Delta u = 0$$

and satisfies the conditions

$$u(0, y) = u(\pi, y) = 0, \quad u(\xi, y) = f(y), \qquad 0 \leq x \leq \pi, \quad 0 \leq y \leq a.$$

11. The inverse problems of tomography are the mathematical problems which deals with interpreting the tomography data. In medicine and in industry the most frequently encountered is X–ray tomography. Interpreting tomography data in medicine tomography is connected with the Radon transformation. The corresponded mathematical problem is as follows.

Let $u(x, y)$ be a continuous finite function with the carrier disposed in a finite domain D. We may suppose that D is the unit disk: $D = \{(x, y) \mid x^2 + y^2 < 1\}$. The Radon transformation of the function $u(x, y)$ is the following function

$$f(x, y, \alpha) = \int_{-\infty}^{\infty} u(x + s\cos\alpha,\ y + s\sin\alpha)\, ds. \tag{9}$$

Equality (9) is considered as a linear operator equation relatively the function $u(x, y)$.

We need to find the function $u(x, y)$ given the function $f(x, y, \alpha)$.

In the mathematical model of X-ray tomography the function $u(x, y)$ is connected with the coefficient of absorption of X-rays. In medicine tomography it is appropriate to consider the desired function as an arbitrary continuous or piecewise continuous function since the distributions of absorption coefficients may rather differ for various people. Moreover, even if we take the concrete man, then the distribution of the absorption coefficient may vary essentially on whether this man stays or lies.

In the industrial tomography, another situation is of an interest. The distribution of the absorption coefficient mostly has the standard form. The goal of tomography research is to find the deviation from the norm, i.e., the defects of the product. If the product is obtained by means of casting, then inside this product there may be cavities and gaps. Thus, we obtain the following mathematical problem: in equation (9), the function $u(x, y)$ is equal to unity in a certain domain $D_0 \subset D$ and outside D_0, it is equal to zero:

$$u(x, y) = \begin{cases} 1, & (x, y) \in D_0 \subset D, \\ 0, & (x, y) \overline{\in} D_0. \end{cases}$$

The boundary of D_0 is a piecewise smooth curve. Since, in this setting, the function $u(x, y)$ is defined by two functions dependent on the one variable, it is natural to suggest that the solution of equation (9) can be obtained by the information which is less complete than the information usually set in the general case relatively the right–hand side $f(x, y, \alpha)$. We consider the problems connected with the solution of (7) in the case when the values of $f(x, y, \alpha)$ are known for two values of the parameter α. So, we suppose that the two functions are known

$$f_1(x, y) = f(x, y, 0), \qquad f_2(x, y) = f(x, y, \alpha)$$

(see Figure 9).

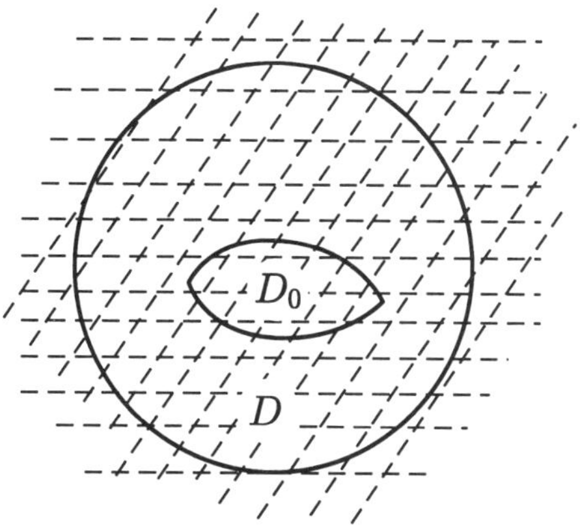

Figure 9

The inverse problem formulated below is analogous to the well-known inverse problem of potential theory which came to the attention of many scientists [11, 6, 12].

Let D be a domain considered in Theorem 3, Section 5:

$$D = \{(x, y) \mid f_1(x) < y < f_2(x),\ x \in (0,1)\},$$

where the functions $f_k(x)$ satisfy conditions 1), 2), 3) mentioned in the formulation of Theorem 4.

Evidently, that if the functions $\varphi_k(y)$, $k = 1, 2$, are inverse relatively the functions $f_k(x)$, i. e.,

$$\varphi_k[f_k(x)] = x,$$

then

$$D = \{(x, y) \mid \varphi_2(y) < x < \varphi_1(y),\ y \in (0,1)\}.$$

Denote

$$u(x) = f_2(x) - f_1(x), \qquad v(y) = \varphi_1(y) - f_2(y).$$

Inverse problem. Find the functions $f_k(x)$ or $\varphi_k(y)$, the functions $u(x)$, $v(y)$ given.

Theorem 6. *The solution to the inverse problem is unique.*

Proof. Let x_0 be an arbitrary point in the interval $(0,1)$:

$$0 < x_0 < 1.$$

Denote

$$f_1(x_0) = y_0, \qquad x_1 = y_0 - v(y_0),$$

then

$$f_2(x_1) = y_0, \qquad f_1(x_1) = y_0 - u(x_1).$$

We consider the sequence

$$\begin{aligned} x_{k+1} &= x_k - v(y_k), \\ y_{k+1} &= y_k - u(x_{k+1}). \end{aligned} \tag{10}$$

Evidently, we have

$$\begin{gathered} x_{k+1} < x_k, \qquad y_{k+1} < y_k; \\ \lim_{n\to\infty} x_n = \lim_{n\to\infty} y_n = 0. \end{gathered}$$

Relations (10) may be considered as the functional equation relatively the value

$$f_1(x_0) = y_0.$$

The solution of this functional equation is unique, as it is easy to show, whence the statement of theorem follows. □

12. Inverse problem for regular domains.

We confine ourselves by the inverse problems considered for the domains satisfying the conditions of Theorem 5.

Let $\varphi_1(x)$, $\varphi_2(x)$, $\psi_1(y)$, $\psi_2(y)$ be continuously differentiable functions defined in the segments $[a_1, a_2]$, $[b_1, b_3]$ so that

1° $\varphi_1(x) < \varphi_2(x)$, $x \in (a_1, a_3)$; $\psi_1(y) < \psi_2(y)$, $y \in (b_1, b_3)$;

2° $\varphi_1(a_1) = \varphi_2(a_1)$, $\varphi_1(a_3) = \varphi_2(a_3)$; $\psi_1(b_1) = \psi_2(b_1)$, $\psi_1(b_3) = \psi_2(b_3)$;

3° $\varphi_1^1(x) < 0$, $x \in (a_1, a_2)$; $\varphi_1^1(x) > 0$, $x \in (a_2, a_3)$; $\varphi_2^1(x) > 0$, $x \in (a_1, a_2)$; $\varphi_2^1(x) < 0$, $x \in (a_2, a_3)$; $\psi_1^1(y) > 0$, $y \in (b_1, b_2)$; $\psi_1^1(y) < 0$, $y \in (b_2, b_3)$; $\psi_2^1(y) < 0$, $y \in (b_1, b_2)$; $\psi_2^1(y) > 0$, $y \in (b_2, b_3)$;

4° $\varphi_1[\psi_1(y)] = y$, $\varphi_2[\psi_2(y)] = y$.

Denote by D the domain

$$D = \{(x, y) \mid \varphi_1(x) < y < \varphi_2(x)\} = \{(x, y) \mid \psi_1(y) < x < \psi_2(y)\}.$$

We suppose also that the domain D satisfies the condition of Theorem 5 and denote by $u(x)$ and $v(y)$ the following functions

$$u(x) = \varphi_2(x) - \varphi_1(x), \qquad v(y) = \psi_2(y) - \psi_1(y). \tag{11}$$

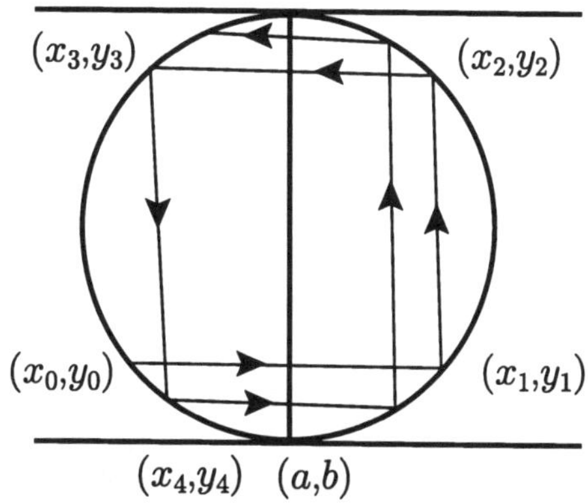

Figure 10

Inverse problem. Find the functions $\varphi_1(x)$, $\varphi_2(x)$, $\psi_1(y)$, $\psi_2(y)$, given the functions $u(x)$, $v(y)$.

The points (a_1, b_2), (a_2, b_3), (a_3, b_2), (a_2, b_1) are the vertices of the domain D.

From (11) it follows that

$$b_3 = b_1 + u(a_2), \qquad a_3 = a_1 + v(b_2).$$

It is easy to see that the numbers a_k, b_k, $k = 1, 2, 3$ are defined uniquely by the functions $u(x)$, $v(y)$.

Really, we have

$$u(x) = 0,\ x < a_1,\ x > a_3, \quad u(x) > 0,\ x \in (a_1,\ a_3), \quad u(a_2) = \max u(x),$$
$$v(y) = 0,\ y < b_1,\ y > b_3, \quad v(y) > 0,\ y \in (b_1,\ b_3), \quad v(b_2) = \max v(y).$$

Let (x_0, y_0) be a certain point in the boundary of D. Let, for example,

$$y_0 = \varphi_1(x_0), \qquad a_1 < x_0 < a_2.$$

Then the points (x_k, y_k) defined by the equalities

$$\begin{aligned} x_{4k+1} &= x_{4k} + v(y_{4k}), & y_{4k+1} &= y_{4k}, \\ x_{4k+2} &= x_{4k+1}, & y_{4k+2} &= y_{4k+1} + u(x_{4k+1}), \\ x_{4k+3} &= x_{4k+2} - v(y_{4k+2}), & y_{4k+3} &= y_{4k+2}, \\ x_{4(k+1)} &= x_{4k+3}, & y_{4(k+1)} &= y_{4k+3} - u(x_{4k+3}) \end{aligned} \tag{12}$$

(see Figure 10) lie also in the boundary of the domain D.

From the supposition that the domain D satisfies the condition of Theorem 5 it follows that

$$\lim_{k\to\infty} x_{2k} = a_2, \qquad \lim_{k\to\infty} y_{2k} = b_1, \qquad \lim_{k\to\infty} y_{2k+1} = b_3. \tag{13}$$

Equalities (12), (13) are the equations for determining $\varphi_1(x_0)$. It is easy to see (taking into account the properties of the functions $\varphi_k(x)$) that the solution of this equation is unique.

So, we had established the following theorem.

Theorem 7. *Solution of the inverse problem formulated above is unique.*

13. Inverse problem for the domains with the a transitive boundary.

Let D be a bounded domain convex relatively the axes (x) and (y) whose boundary is continuously differentiable. Denote the vertices of this domain by (a_1, b_1), (a_2, b_2), (a_3, b_3), (a_4, b_4), $a_1 < a_2 < a_3 < a_4$; $b_2 < b_1 < b_4 < b_3$. Then there exist the functions $f_1(x)$, $f_2(x)$, $g_1(y)$, $g_2(y)$ continuously differentiable in the intervals $x \in (a_1, a_4)$, $y \in (b_2, b_3)$ such that

$$\begin{aligned} D &= \{(x,y) \mid f_1(x) < y < f_2(x),\ \ x \in (a_1, a_4)\} \\ &= \{(x,y) \mid g_1(y) < x < g_2(y),\ \ y \in (b_2, b_3)\}; \end{aligned}$$

1) $f_1(a_1) = f_2(a_1)$, $f_1(a_4) = f_2(a_4)$; $g_1(b_2) = g_2(b_2)$, $g_1(b_3) = g_2(b_3)$;

2) $f_1(x) = f_2(x)$, $x \in (a_1, a_4)$; $g_1(y) = g_2(y)$, $y \in (b_2, b_3)$;

3) $f_1'(x) < 0$, $x \in (a_1, a_2)$; $f_1'(x) > 0$, $x \in (a_2, a_4)$; $f_2'(x) > 0$, $x \in (a_1, a_3)$; $f_2'(x) < 0$, $x \in (a_3, a_4)$;

4) $g_1[f_1(x)] = x$, $x \in (a_1, a_2)$; $g_1[f_2(x)] = x$, $x \in (a_2, a_3)$; $g_2[f_1(x)] = x$, $x \in (a_2, a_4)$; $g_2[f_2(x)] = x$, $x \in (a_3, a_4)$;

We denote by $u(x)$, $v(y)$ the functions

$$u(x) = f_2(x) - f_1(x), \qquad v(y) = g_2(y) - g_1(y). \tag{14}$$

Inverse problem. Given the functions $u(x)$, $v(y)$, find the functions $f_1(x)$, $f_2(x)$, $g_1(y)$, $g_2(y)$.

We assume, in addition, that the curve C that bounded the domain D is transitive and the vertices of C are known.

Theorem 8. *The solution of this inverse problem is unique.*

Proof. We consider the left vertex (a_1, b_1) of the curve C. By (14), the point $(x_1, b_1) \in C$, where $x_1 = a_1 + v(b_1)$. Further, we have $(x_1, y_1) \in C$, where $y_1 = b_1 + u(x_1)$.

Consider the sequence of points (x_k, y_k), (x_{k+1}, y_k), $k = 1, 2, \ldots$ defined as follows:

$$\begin{aligned} x_{k+1} &= x_k \pm v(y_k), \\ y_{k+1} &= y_k \pm u(x_{k+1}). \end{aligned} \tag{15}$$

The signs before $v(y_k)$, $u(x_k)$ in (15) are defined dependently on the mutual disposition of the points (x_k, y_k), (x_{k+1}, y_k) and the vertices of the curve C.

Since, by our assumption, the curve C is transitive, the points (x_k, y_k), (x_{k+1}, y_k) form the everywhere dense set in the curve C, whence the theorem follows. □

REFERENCES

1. R. A. Aleksandryan, On a Sobolev problem for special equations with partial derivatives of the fourth order. *Dokl. Akad. Nauk SSSR* (1950), **73**, No. 4 (in Russian).

2. T. I. Zelenyak, *Selected Questions of Qualitative Theory of Equations with Partial Derivatives.* Novosibirsk Gos. University, 1970 (in Russian).

3. V. K. Imanaliev, P. S. Pankov and O. G. Brender, The method of subsequent differentiating for construction of generalized solution of the integral Fredholm equation of the first kind. In: *Research of Integro–Differential Equations Akad. Nauk KSSR*, Frunze, 1983 (in Russian).

4. R. Courant, *Equations with Partial Derivatives.* Mir, Moscow, 1964 (in Russian).

5. M. M. Lavrent'ev, *On Certain Ill-Posed Problems of Mathematical Physics.* Sib. Otdel Akad. Nauk SSSR, 1962 (in Russian).

6. M. M. Lavrent'ev, V. G. Romanov and S. P. Shishatskii, *Ill-Posed Problems of Mathematical Physics and Analysis.* Nauka, Moscow, 1980 (in Russian).

7. M. M. Lavrent'ev, On a boundary-value problem for hyperbolic system. In: *Matem. Zbornik* (1956), 38(60) (in Russian).

8. R. Lattes and J.-L. Lions, *Quasiinversion Method and Its Applications.* Mir, Moscow, 1970 (in Russian).

9. M. A. Lavrent'ev and B. V. Shabat, *Problems of Hydrodynamics and Their Mathematical Models.* Nauka, Moscow, 1977 (in Russian).

10. P. S. Novikov, On uniqueness of inverse problem of potential theory. *Dokl. Akad. Nauk SSSR* (1938), **18**, No. 3 (in Russian).

11. A. I. Prilepko, On inverse problems of potential theory. *Differential Equations* (1967) **3**, No. 1 (in Russian).

12. I. G. Petrovskii, *Lections on Equations with Partial Derivatives.* Fizmatgiz, Moscow, 1961 (in Russian).

13. S. L. Sobolev, *Equations of Mathematical Physics.* Nauka, Moscow, 1966 (in Russian).

14. V. H. Strakhov, On solution of ill-posed problems of magneto and gravimetry represented by integral equations of convolution type. *Izv. Akad. Nauk SSSR. Fizika Zemli* (1967) **4** (in Russian).

15. A. N. Tikhonov, On stability of inverse problems. *Dokl. Akad. Nauk SSSR* (1943) **39**, No. 5 (in Russian).

16. A. N. Tikhonov and V. Ya. Arsenin, *Solutions of Ill–Posed Problems.* Wiley, New York, 1977.

17. A. N. Tikhonov, On solution of ill-posed problems and regularization method. *Dokl. Akad. Nauk SSSR* (1963) **153**, No. 1 (in Russian).

18. B. V. Shabat, On the analog of Riemann's theorem for hyperbolic system of differential equations. *Uspekhi Math. Nauk* (1956) **11**, 3(69) (in Russian).

19. B. V. Shabat, On hyperbolic quasiconformal mappings. In: *Certain Problems of Mathematics and Mechanics.* Nauka, Leningrad, 1970 (in Russian).

20. Sh. Yarmuhamedov, On analytic extension of holomorphic vector by its values on the part of boundary. *Izv. Akad. Nauk Uz. SSR, Ser. Fiz.-Mat.* (1980) , No. 6 (in Russian).

21. J. Douglas, *Numerical Method for Analytic Continuation, Boundary Problems Different Equations.* Madison Univ., Wisconsin Press, 1960.

22. J. Hadamard, Sur les problemes aux derivees partielles et leur signification physique. *Bull. Univ. Princeton* (1902) **13**.

23. J. Hadamard, *Le Probleme de Cauchy et les Equations aux Derivees Partielles Lineaires Hyperboliques.* Paris, Herman, 1932.

24. F. John, The Dirichlet problem for a hyperbolic equation. *Amer. J. Math.* (1941) **63**.

25. C. Pucci, Discussione del problema di Cauchy pur le equazioni di tipo elliptico. *Ann. Mat Pura ed Appl.* (1958).

26. M. Yamamoto, Conditional stability in determination of force terms of heat equations in rectangle. *Math. and Comp. Modelling* (1993) **18**.

Ill-Posed and Non-Classical Problems of
Mathematical Physics and Analysis, Samarkand, 2000, pp. 103–124
M. M. Lavrent'ev and S. I. Kabanikhin (Eds)

Systems of linear integral equations of Volterra type with singular and super-singular kernels

N. RAJABOV

Abstract — In this paper linear systems of integral equations of Volterra type with the left and right fixed singular and super-singular points are researched. Depending on disposition of singular points, on the singularity order, and on the roots of the correspondent characteristic equations, the general representation of the solution set is obtained. The cases when the correspondent homogeneous systems of integral equations have nontrivial solutions and when there exist no solutions except the trivial solution are considered. In the case when the homogeneous system has the trivial solution only, the correspondent inhomogeneous system has a unique solution. In the other cases the general solutions of the inhomogeneous systems depend on several arbitrary constants. In some particular cases the solutions of these systems of integral equations are found explicitly. The behaviour of solutions in the neighborhood of singular points is investigated. Note that the problem of determining the collection of continuous solutions of systems of ordinary differential equations with fixed boundary singular and super-singular points is reduced to consideration of such systems of integral equations [1, 2].

1. THE CASE WHEN THE KERNELS ARE THE FUNCTIONS OF VARIABLE INTEGRATION

Let $\Gamma = \{a < x < b\} \subset R^1$. We consider the following system of integral equations

$$y_j(x) + \sum_{k=1}^{n} \int_a^x (t-a)^{-\alpha} a_{jk}(t) y_k(t)\, dt = f_j(x), \quad 1 \le j \le n, \tag{1.1}$$

where $a_{jk}(x)$, $f_j(x)$, $(1 \le j,\ k \le n)$ are known continuous functions of the points from $\overline{\Gamma}$, $n \in N$, and $\alpha = \text{const} > 0$.

We shall seek the solution of system (1.1) in the class of functions continuous in $\overline{\Gamma}$: $y_j(x)$, $(1 \le j \le n)$, which have the proximity order in the point $x = a$ higher than $\alpha - 1$:

$$y_j(x) = o[(x-a)^{\gamma_j}], \quad \gamma_j > \alpha - 1 \quad \text{for} \quad x \to a.$$

The scalar equation of type (1.1) was investigated in [3, 7].

In this section the general solution of system (1.1) is obtained depending on α ($\alpha < 1$; $\alpha = 1$; $\alpha > 1$) and the roots of characteristic equation

$$\Delta(\lambda) \equiv \det \|a_{jk}(a) - \lambda \delta_{jk}\| = 0. \tag{1.2}$$

In the case when $a_{jk}(x) = a_{jk} = \text{const}$, the solution of system (1.1) is found explicitly.

When the roots of characteristic equation (1.2) are real and distinct the following statements hold.

Theorem 1.1. *Suppose that $\alpha > 1$ in system (1.1) and the roots of characteristic equation (1.2) λ_j $(1 \le j \le n)$ are real and distinct; $\lambda_j < 0$; $(1 \le j \le n)$. Let, besides, $f_j(x)$, $a_{jk}(x) \in C(\overline{\Gamma})$; $(1 \le j,\ k \le n)$ and satisfy the following conditions for $x \to a$:*

$$f_j(x) = o\Big[\exp\Big(\Big(\sum_{m=1}^{n} \lambda_m\Big)\omega_\alpha(x)\Big) \cdot (x-a)^{\gamma_1}\Big], \quad \gamma_1 > \alpha - 1, \tag{1.3}$$

$$a_{jk}(x) - a_{jk}(a) = o\Big[\exp\Big(\Big(\sum_{m=1}^{n} \lambda_m\Big)\omega_\alpha(x)\Big) \cdot (x-a)^{\gamma_1}\Big], \tag{1.4}$$

$$\gamma_2 > \alpha - 1, \quad \omega_\alpha(x) = [(\alpha-1)(x-a)^{\alpha-1}]^{-1}, \quad 1 \le j,\ k \le n.$$

Then homogeneous system (1.1) has n solutions linearly independent. Inhomogeneous system (1.1) is solvable in the class $C(\overline{\Gamma})$ and its general solution has n arbitrary constants. In this case the general solution of system (1.1) is expressed via the resolvents of the system of Volterra equations of the second kind with a weak singularity in the point $x = a$:

$$y_j(x) = T_j^{(\alpha)}[c_1, c_2, \ldots, c_n, f_1(x), f_2(x), \ldots, f_n(x)]$$
$$+ \sum_{k=1}^{n} \int_a^x \Gamma_{jk}^{(\alpha)}(x,t) T_k^{(\alpha)}[c_1, c_2, \ldots, c_n, f_1(t), \ldots, f_n(t)]\, dt,$$

where c_j $(1 \le j \le n)$ are arbitrary constants.

In this representation $\Gamma^{(\alpha)}_{jk}(x,t)$ are the resolvents of the following system of integral equations

$$y_j(x) - \sum_{l=1}^{n} \int_a^x K^{(\alpha)}_{jl}(x,t) y_l(t)\,\mathrm{d}t = T^{(\alpha)}_j[c_1, c_2, \dots, c_n, f_1(x), \dots, f_n(x)],$$
$$1 \le j \le n,$$

$$K^{(\alpha)}_{jl}(x,t) = \sum_{k=1}^{n} c_{kj} \sum_{m=1}^{n} \frac{\Delta_{mk}}{\Delta}\Big(\frac{a_{ml}(t) - a_{ml}(a)}{(t-a)^{\alpha}}\Big) \exp\left[\lambda_k(\omega_\alpha(x) - \omega_\alpha(t))\right],$$

$$T^{(\alpha)}_j[c_1, c_2, \dots, c_n, f_1(x), \dots, f_n(x)] = \sum_{k=1}^{n} c_{kj} \exp[\alpha_k \omega_\alpha(x)] c_k$$
$$\not\equiv \sum_{k=1}^{n} c_{kj} \sum_{m=1}^{n} \frac{\Delta_{mk}}{\Delta}\Big[f_m(x) - \alpha_k \int_a^x \exp\left[\lambda_k(\omega_\alpha(x) - \omega_\alpha(t))\right] f_m(t)(t-a)^{-\alpha}\,\mathrm{d}t\Big],$$
$$1 \le j \le n,$$

where $\Delta = \det \|c_{kj}\|$, Δ_{mk} are the cofactors of c_{mk}; c_{kj} are the solutions of the system

$$\sum_{k=1}^{n} a_{jk}(a) c_k - \lambda c_j = 0 \qquad (1 \le j \le n)$$

correspondent to the values $\lambda = \lambda_k$ $(1 \le k \le n)$.

Theorem 1.2. *Let $\lambda = 1$ in system (1.1) and the roots of characteristic equation (1.2) λ_j $(1 \le j \le n)$ be real and distinct; $\lambda_j < 0$ $(1 \le j \le n)$. Let, moreover, $f_j(x)$, $a_{jk}(x) \in C(\overline{\Gamma})$ $(1 \le j, k \le n)$ and satisfy the following conditions for $x \to a$*

$$f_j(x) = o\big[(x-a)^{-\sum_{m=1}^{n} \lambda_m + \gamma_3}\big], \quad \gamma_3 > 0,$$
$$a_{jk}(x) - a_{jk}(a) = o\big[(x-a)^{-\sum_{m=1}^{n} \lambda_m + \gamma_4}\big], \quad \gamma_4 > 0, \quad 1 \le j,\ k \le n.$$

Then homogeneous system (1.1) has n linearly independent solutions. Inhomogeneous system (1.1) is solvable and its general solution belongs to the class $C(\overline{\Gamma})$ and contains n arbitrary constants. General solution of system (1.1) is expressed via the resolvents of the system of integral Volterra equations of the second kind with the weak singular point $x = a$:

$$y_j(x) = T^{(1)}_j[c_1, c_2, \dots, c_n, f_1(x), f_2(x), \dots, f_n(x)]$$
$$+ \sum_{k=1}^{n} \int_a^x \Gamma_{jk}(x,t) T_k[c_1, c_2, \dots, c_n, f_1(t), \dots, f_n(t)]\,\mathrm{d}t,$$

where c_j $(1 \le j \le n)$ are arbitrary numbers.

Here we denote by $\Gamma_{jk}(x,t)$ the resolvents of the following system of Volterra equations of the second kind with the weak singularity

$$y_j(x) - \sum_{l=1}^{n} \int_a^x K_{jl}(x,t) y_k(t)\,\mathrm{d}t = T_j^{(1)}[c_1, c_2, \ldots, c_n, f_1(x), \ldots, f_n(x)],$$
$$1 \le j \le n,$$

$$K_{jl}(x,t) = \sum_{k=1}^{n} c_{kj} \sum_{m=1}^{n} \frac{\Delta_{mk}}{\Delta} \Big(\frac{a_{ml}(t) - a_{ml}(a)}{(t-a)}\Big) \Big(\frac{t-a}{x-a}\Big)^{\lambda_k},$$

$$T_j^{(1)}[c_1, c_2, \ldots, c_n, f_1(x), \ldots, f_n(x)] = \sum_{k=1}^{n} c_{kj} (x-a)^{-\lambda_k}$$
$$+ \sum_{m=1}^{n} \frac{\Delta_{mk}}{\Delta} \Big[f_m(x) - \lambda_k \int_a^x \Big(\frac{t-a}{x-a}\Big)^{\lambda_k} \frac{f_m(t)}{t-a}\Big]\,\mathrm{d}t, \quad 1 \le j \le n$$

Now, we shall give the proofs of these theorems.

Proof of Theorem 1.1. Suppose that $\alpha > 1$ in system (1.1). Adding and subtracting the numbers $a_{jk}(a)$ from the integrands, we rewrite the system as follows:

$$y_j(x) + \sum_{k=1}^{n} a_{jk}(a) \int_a^x (t-a)^{-\alpha} y_k(t)\,\mathrm{d}t = F_j^{(\alpha)}(x), \quad 1 \le j \le n, \tag{1.5}$$

where

$$F_j^{(\alpha)}(x) = f_j(x) - \sum_{k=1}^{n} \int_a^x (t-a)^{-\alpha} [a_{jk}(t) - a_{jk}(a)] y_k(t)\,\mathrm{d}t, \quad 1 \le j \le n. \tag{1.6}$$

Since the functions $F_j^{(\alpha)}(x)$ $(1 \le j \le n)$ are known, we find at first the general solution of system (1.5).

In the case when the functions $F_j^{(\alpha)}(x)$ are known $(a_{jk}(x) = a_{jk}(a))$ we shall call the system (1.5) by the model system or by characteristic system of integral Volterra equations with the super-singular $(\alpha > 0)$ left fixed point in the kernel.

Assume that system (1.5) has the solutions, these solutions are differentiable and $F_j^{(\alpha)}(x) \in C'(\Gamma)$ $(1 \le j \le n)$. Differentiating both sides of (1.5) we obtain the following linear system of ordinary differential equations with the left boundary singular ($\alpha = 1$) or super-singular ($\alpha > 1$) point. The theory of such systems

$$y_j'(x) + (x-a)^{-\alpha} \sum_{k=1}^{n} a_{jk}(a) y_k(x) = F_j'(x), \quad 1 \le j \le n. \tag{1.7}$$

is well developed in [1]. It is known [1] that this system has also the solutions which are unbounded in the point $x = a$. We need to find the continuous solution of this system for which $x = a$ is the zero of the order higher than $\alpha - 1$.

Firstly, as in [1], we find the solution of the homogeneous system

$$y_j'(x) + (x-a)^{-\alpha} \sum_{k=1}^{n} a_{jk}(a) y_k(x) = 0. \tag{1.8}$$

The solution of system (1.8) we shall seek in the form (see [1])

$$y_j(x) = c_j \exp[\lambda \omega_\alpha(x)], \qquad \omega_\alpha(x) = [(\alpha-1)(x-a)^{\alpha-1}]^{-1}$$
$$(1 \le j \le n). \tag{1.9}$$

Acting similarly as in the theory of ordinary differential equations with constant coefficients we pass to the following algebraic system

$$\sum_{k=1}^{n} a_{jk}(a) c_k - \lambda c_j = 0 \qquad (1 \le j \le n). \tag{1.10}$$

The determinant of this system has the following form

$$\Delta(\lambda) = \det \|a_{jk}(a) - \lambda \delta_{jk}\| = 0. \tag{1.11}$$

Suppose that the roots of algebraic equation (1.11) are real and distinct where $\lambda_j < 0$ $(1 \le j \le n)$. The following fundamental system of solutions corresponds to these roots

$$y_{kj}(x) = c_{kj} \exp[\lambda_k \omega_\alpha(x)], \quad 1 \le j, \;\; k \le n,$$

where c_{kj} solve system (1.10) for the values $\lambda = \lambda_j$ $(1 \le j \le n)$.

Then the general solution of system (1.8) from the class $C'(\Gamma) \cap C(\overline{\Gamma})$ is given by the formula

$$y_j(x) = \sum_{k=1}^{n} a_{kj} c_k \exp[\lambda_k \omega_\alpha(x)], \quad \lambda_k < 0 \; (1 \le k \le n), \quad 1 \le j \le n,$$

where c_k $(1 \le k \le n)$ are arbitrary constants. By the conditions of Theorem 1.1, $\lambda_k < 0$ $(1 \le k \le n)$; therefore, $y_j(x) \in C'(\overline{\Gamma})$ and in the point $x = a$ they vanish.

Using the method of variation of parameters, we obtain that if the solution of inhomogeneous system (1.7) exists, then it has the form

$$y_j(x) = \sum_{k=1}^{n} \exp\,[\lambda_k \omega_\alpha(x)] c_{kj} \Big[c_k + \sum_{m=1}^{n} \frac{\Delta_{mk}}{\Delta} \int_a^x \exp\,[-\lambda_k \omega_\alpha(t)] F'_m(t)\,\mathrm{d}t\Big],$$
$$1 \le j \le n,$$

where $\Delta = \det \|c_{kj}\|$, Δ_{mk} are the cofactors of c_{mk}, c_k $(1 \le k \le n)$ are arbitrary constants.

Integrating by terms the integrals which stand in the right-hand side of this equality and setting

$$\exp\,(-\lambda_m \omega_\alpha(t)) F_m(t)\big|_{t=a} = 0 \quad (1 \le m \le n), \tag{1.12}$$

we obtain the following equalities

$$\begin{aligned} y_j(x) &= \sum_{k=1}^{n} c_{kj} \exp\,[\lambda_k \omega_\alpha(x)] c_k + \sum_{k=1}^{n} c_{kj} \sum_{m=1}^{n} \frac{\Delta_{mk}}{\Delta} \Big[F_m^{(\alpha)}(x) \\ &\quad - \lambda_k \int_a^x \exp\,[\lambda_k(\omega_\alpha(x) - \omega_\alpha(t))](t-a)^{-\alpha} F_m^{(\alpha)}(t)\,\mathrm{d}t\Big] \\ &\equiv T_j^\alpha[c_1, \dots, c_n; F_1^{(\alpha)}(x), \dots, F_n^{(\alpha)}(x)], \quad 1 \le j \le n. \end{aligned} \tag{1.13}$$

These solutions of system (1.5) were obtained in supposition that the desired functions $y_j(x)$ $(1 \le j \le n)$ and the function $F_j^{(\alpha)}(x)$ $(1 \le j \le n)$ from this system are differentiable and $F_j^{(\alpha)}(x)$ $(1 \le j \le n)$ satisfy the conditions of type (1.3). By immediate computation we can verify that the functions $y_j(x)$ $(1 \le j \le n)$ defined by (1.13) satisfy the system of integral equations (1.5) for each $F_j^{(\alpha)}(x) \in C(\overline{\Gamma})$ $(1 \le j \le n)$ satisfying conditions (1.12). □

Thus, we have proved the following lemma.

Lemma 1.1. *Let in system (1.5) the numbers $a_{jk}(a)$ $(1 \le j,\ k \le n)$ be such that the roots of characteristic equation (1.2) λ_j $(1 \le j \le n)$ are real and distinct and $\lambda_j < 0$ $(1 \le j \le n)$. Let, moreover $F_j^{(\alpha)}(x) \in C'(\overline{\Gamma})$ $(1 \le j \le n)$, they vanish in the point $x = a$ and in the neighborhood of $x = a$ the following asymptotic formulas hold*

$$F_j^{(\alpha)}(x) = o[\exp\,(-|\lambda|\omega_\alpha(x))(x-a)^\delta], \quad \delta > \alpha - 1, \quad 1 \le j \le n. \tag{1.14}$$

Then system (1.5) is solvable and its general solution contains n arbitrary constants and is given by formula (1.13).

Suppose now that the functions $F_j^{(\alpha)}(x)$ $(1 \le j \le n)$ have form (1.6); $y_j(x)$ $(1 \le j \le n)$ are unknown. By immediate computation we can verify that if the functions $f_j(x)$ $(1 \le j \le n)$, $a_{jk}(x) - a_{jk}(a)$ $(1 \le j,\ k \le n)$ satisfy conditions (1.3) and (1.4) of Theorem 1.1 respectively, $y_j(x) \in C(\overline{\Gamma})$ $(1 \le j \le n)$ and vanish for $x \to a$, then the functions $F_j^{\alpha}(x)$ $(1 \le j \le n)$ satisfy conditions (1.14) for $x \to a$. Act by the integral operator which is inverse to the integral operator (1.5)

$$T^{\alpha} \equiv \begin{pmatrix} T_1^{\alpha} \\ \vdots \\ T_n^{\alpha} \end{pmatrix}$$

on both sides of equality (1.13). As a result we obtain the following system of linear integral Volterra equations with the weak singular point $t = a$

$$y_j(x) - \sum_{l=1}^{n} \int_a^x K_{jl}(x,t) y_l(t)\, dt = T_j^{\alpha}[c_1, \dots, c_n; f_1(x), \dots, f_n(x)],$$
$$1 \le j \le n, \tag{1.15}$$

where

$$K_{jl}(x,t) = \sum_{k=1}^{n} c_{kj} \sum_{m=1}^{n} \frac{\Delta_{mk}}{\Delta} \Big(\frac{a_{ml}(t) - a_{ml}(a)}{(t-a)^{\alpha}} \Big) \exp\left[\lambda_k(\omega_{\alpha}(x) - \omega_{\alpha}(t))\right],$$
$$1 \le k \le n, \qquad \lambda_k < 0. \tag{1.16}$$

Solving system (1.15) we find

$$y_j(x) = T_j^{\alpha}[c_1, \dots, c_n; f_1(x), \dots, f_n(x)]$$
$$+ \sum_{l=1}^{n} \int_a^x \Gamma_{jl}(x,t) T_l^{\alpha}[c_1, \dots, c_n; f_1(t), \dots, f_n(t)]\, dt, \tag{1.17}$$

where $\Gamma_{jl}(x,t)$ $(1 \le j,\ l \le n)$ are the resolvents of system (1.15).

Proof of Theorem 1.2. Let $\alpha = 1$. In this case, for system (1.1), we obtain the following characteristic system

$$y_j(x) + \sum_{k=1}^{n} a_{jk}(a) \int_a^x \frac{y_k(t)}{t-a}\, dt = F_j^{(1)}(x), \quad 1 \le j \le n, \tag{1.18}$$

where

$$F_j^{(1)}(x) = f_j(x) - \sum_{k=1}^{n} \int_a^x \frac{a_{jk}(t) - a_{jk}(a)}{t-a} y_k(t)\,\mathrm{d}t, \quad 1 \le j \le n. \tag{1.19}$$

The solution of homogeneous system (1.18) we shall seek in the form

$$y_j(x) = c_j(x-a)^{-\lambda} \quad (1 \le j \le n).$$

Acting following the scheme of the proof of Theorem 1.1 we see that if the functions $F_j^{(1)}(x)$ $(1 \le j \le n)$ for $x \to a$ have the behaviour

$$F_j^{(1)}(x) = o\Big[(x-a)^{\sum_{k=1}^{n} |\lambda_k| + \delta}\Big], \quad \delta > 0, \quad (1 \le j \le n), \tag{1.20}$$

the roots of characteristic equation (1.2) are real and distinct and $\lambda_k < 0$ $(1 \le j \le k)$, then the solution of system (1.18) from the class $C(\overline{\Gamma})$ is given by the formula

$$\begin{aligned} y_j(x) &= \sum_{k=1}^{n} c_{kj}(x-a)^{-\lambda_k} \\ &\quad + \sum_{m=1}^{n} \frac{\Delta_{mk}}{\Delta} \Big[F_m^{(1)}(x) - \lambda_k \int_a^x \Big(\frac{t-a}{x-a}\Big)^{\lambda_k} \frac{F_m^{(1)}(t)}{t-a}\,\mathrm{d}t\Big] \\ &\equiv T_j^{(1)}[c_1, \ldots, c_n, F_1^{(1)}(x), \ldots, F_n^{(1)}(x)], \quad (1 \le j \le n). \end{aligned} \tag{1.21}$$

□

From these considerations the below lemma follows.

Lemma 1.2. *Suppose in equation (1.18) the functions $F_j^{(1)}(x) \in C(\overline{\Gamma})$ vanish for $x \to a$ and their behaviour is defined in asymptotic formula (1.20). Suppose also that the roots of equation (1.2) are real and distinct, $\lambda_k < 0$ $(1 \le k \le n)$. Then system (1.18) is solvable and its general solution from the class $C(\overline{\Gamma})$ contains n arbitrary constants and is given by formula (1.21).*

Substituting the values $F_m^{(1)}(x)$ $(1 \le m \le n)$ into representation (1.21) and repeating the scheme of the proof of Theorem 1.1 when determining $y_j(x)$ $(1 \le j \le n)$, we come to the following system of integral equations

$$y_j(x) - \sum_{l=1}^{n} \int_a^x K_{jl}(x,t) y_l(t)\,\mathrm{d}t = T_j^{(1)}[c_1, \ldots, c_n, f_1(x), \ldots f_n(x)], \quad 1 \le j \le n, \tag{1.22}$$

where

$$K_{jl}(x,t) = \sum_{k=1}^{n} c_{kj} \sum_{m=1}^{n} \frac{\Delta_{mk}}{\Delta}\left(\frac{a_{ml}(t) - a_{ml}(a)}{t-a}\right)\left(\frac{t-a}{x-a}\right)^{\lambda_k}. \qquad (1.23)$$

Suppose that the functions $a_{ml}(t) - a_{ml}(a)$ satisfy the conditions of Theorem 1.2 and $\lambda_k < 0$ $(1 \le k \le n)$. Then the kernels $K_{jl}(x,t)$ defined by formula (1.23) will have the weak singularity in the neighborhood of the point $t = a$ and will be continuous for each fixed t with respect to x. If the functions $f_j(x)$ $(1 \le j \le n)$ satisfy the conditions of Theorem 1.2, then the functions $T_j^{(1)}[c_1, \ldots, c_n, f_1(x), \ldots f_n(x)]$ $(1 \le j \le n)$ will be continuous for $\lambda_k < 0$ $(1 \le j \le n)$. From the theory of linear systems of Volterra equations it follows that system (1.22) has a unique solution

$$\begin{aligned} y_j(x) &= T_j^{(1)}[c_1, \ldots, c_n, f_1(x), \ldots f_n(x)] \\ &+ \sum_{l=1}^{n} \int_a^x \Gamma_{jl}(x,t) T_l^{(1)}[c_1, \ldots, c_n, f_1(t), \ldots f_n(t)]\,\mathrm{d}t, \quad 1 \le j \le n, \qquad (1.24) \end{aligned}$$

where $\Gamma_{jl}(x,t)$ are the resolvents of system (1.22).

Corollary 1.1. *In the case when* $a_{jk}(x) = \text{const}$ *and the first conditions of Theorem 1.1 or Theorem 1.2 hold, then the general solution of system* (1.1) *from the class* $C(\overline{\Gamma})$ *is written explicitly by the formula*

$$y_j(x) = T_j^{(\alpha)}[c_1, \ldots, c_n; f_1(x), \ldots f_n(x)], \quad 1 \le j \le n, \quad \alpha \ge 1,$$

where $c_1, c_2, \ldots, c_n$ *are arbitrary constants.*

Corollary 1.2. *Suppose that* $a_{jk}(x) \equiv a_{jk} = \text{const}$, $\alpha \ge 1$, *the roots of characteristic equation* (1.2) *are real and distinct,* $\lambda_j > 0$ $(1 \le j \le n)$, $f_j(x) \in C(\overline{\Gamma})$ $(1 \le j \le n)$ *and satisfy the following conditions for* $x \to a$:

$$f_j(x) = o[(x-a)^{\varepsilon}], \quad 1 \le j \le n, \quad \varepsilon > \alpha - 1.$$

Then homogeneous system (1.1) *has only the trivial solution and the inhomogeneous system* (1.1) *has a unique solution. In this case the general solution of system* (1.1) *from the class* $C(\overline{\Gamma})$ *is given explicitly by the formula*

$$y_j(x) = T_j^{(\alpha)}[0, 0, \ldots, 0; f_1(x), \ldots, f_n(x)], \quad 1 \le j \le n.$$

In the case when $n = 2$, $a_{jk}(x) = a_{jk} = \text{const}$ and the conditions of Theorem 1.1 or of Theorem 1.2 hold, the general solution of system (1.1) from the class $C(\overline{\Gamma})$ is given respectively by the formulas

$$y_j(x) = K_{j,1}^{\lambda_1,\lambda_2}[c_1, c_2, f_1(x), f_2(x)] \equiv \sum_{m=1}^{2} c_m c_{mj}(x-a)^{-\lambda_m}$$
$$+ \Delta_0^{-1}[c_{1j}(c_{22}f_1(x) - c_{21}f_2(x)) + c_{2j}(c_{11}f_2(x) - c_{21}f_1(x))]$$
$$+ \Delta_0^{-1}\int_a^x \Big[\Big(\frac{t-a}{x-a}\Big)^{\lambda_1} c_{1j}\lambda_1(c_{22}f_1(t) - c_{21}f_2(t))$$
$$+ \Big(\frac{t-a}{x-a}\Big)^{\lambda_2} c_{2j}\lambda_2(c_{11}f_2(x) - c_{21}f_1(x)) \Big] \frac{dt}{t-a}, \quad j = 1,2$$

for $\alpha = 1$ and

$$y_j(x) = K_{j,\alpha}^{\lambda_1,\lambda_2}[c_1, c_2, f_1(x), f_2(x)] \equiv \sum_{m=1}^{2} c_m c_{mj} \exp[\lambda_m \omega_\alpha(x)]$$
$$+ \Delta_0^{-1}[c_{1j}(c_{22}f_1(x) - c_{21}f_2(x)) + c_{2j}(c_{11}f_2(x) - c_{12}f_1(x))]$$
$$- \Delta_0^{-1}\int_a^x \Big\{ \exp[\lambda_1(\omega_\alpha(x) - \omega_\alpha(t))]\lambda_1(c_{22}f_1(t) - c_{21}f_2(x)) \cdot c_{1j}$$
$$+ \exp[\lambda_2(\omega_\alpha(x) - \omega_\alpha(t))] \cdot \lambda_2 c_{2j}(c_{11}f_2(t) - c_{12}f_1(t)) \Big\} (t-a)^{-\alpha}\, dt, \ j = 1,2$$

for $\alpha > 1$. Here c_1, c_2 are arbitrary constants, $\Delta_0 = c_{11}\cdot c_{22} - c_{12}\cdot c_{21}$; c_{1j}, c_{2j} $(j = 1,2)$ are the solutions of the algebraic system

$$\sum_{s=1}^{2}(a_{sj} - \delta_{sj}\lambda) = 0, \qquad j = 1,2$$

for $\lambda = \lambda_1$ and $\lambda = \lambda_2$ respectively; $\delta_{ss} = 1$, $\delta_{sj} = 0$ for $s \neq j$.

Further, for simplicity, we shall formulate the results for the case $n = 2$.

In this case $n = 2$, when in system (1.1) the roots of characteristic equation (1.2) are real and equal or the roots are complex conjugate, the following statements hold.

Theorem 1.3. *Suppose in system (1.1) $\alpha > 1$, $n = 2$ and the roots of characteristic equation (1.2) are real and equal: $\lambda_1 = \lambda_2 = \lambda < 0$. Let, moreover, the functions $f_j(x)$, $a_{jk}(x) \in C(\overline{\Gamma})$ $(1 \leq j,\ k \leq 2)$ satisfy for $x \to a$ the following conditions*

$$f_j(x) = o[\exp(-|\lambda|\omega_\alpha(x))(x-a)^{\gamma_5}], \quad \gamma_5 > 2(\alpha - 1), \tag{1.5}$$
$$a_{jk}(x) - a_{jk}(a) = o[\exp(-|\lambda|\omega_\alpha(x))(x-a)^{\gamma_6}], \quad \gamma_6 > 2(\alpha - 1). \tag{1.6}$$

Then homogeneous system (1.1) has two linearly independent solutions. Inhomogeneous system (1.1) is always solvable in the class $C(\overline{\Gamma})$, and its general solution has the two arbitrary constants. The general solution of system (1.1) is expressed by the resolvents of the system of Volterra equations with the weak singularity in the point $t = a$:

$$y_j(x) = M_j[c_1, c_2, f_1(x), f_2(x)] + \sum_{k=1}^{2} \int_a^x \Gamma_{jk}(x,t) M_j[c_1, c_2, f_1(t), f_2(t)]\, dt, \quad 1 \le j \le 2,$$

where c_1 and c_2 are arbitrary constants.

We denote here by $\Gamma_{jk}(x,t)$ the resolvents of the following system of integral equations

$$y_j(x) - \sum_{m=1}^{2} \int_a^x K_{jm}(x,t) y_m(t)\, dt = M_j[c_1, c_2, f_1(x), f_2(x)], \quad 1 \le j \le 2,$$

where $K_{jm}(x,t)$ are the known functions dependent on the coefficients of system (1.1).

Corollary 1.3. *Let $n = 2$, $a_{jk}(x) = a_{jk} = \text{const}$ and condition (1.5) of Theorem 1.3 hold. Then the general solution of system (1.1) from the class $C'(\Gamma) \cap C(\overline{\Gamma})$ is as follows*

$$y_j(x) = M_j[c_1, c_2, f_1(x), f_2(x)], \quad j = 1, 2$$

where

$$\begin{aligned} M_1[c_1, c_2, f_1(x), f_2(x)] &= \exp[\lambda \omega_\alpha(x)](c_1 + \omega_\alpha c_2) \\ &\quad + [(a_{22} - a_{11}) 2^{-1} \omega_\alpha(x) + 2] f_1(x) - a_{12} \omega_\alpha(x) f_2(x) \\ &- \int_a^x \exp[\lambda(\omega_\alpha(x) - \omega_\alpha(t))](t-a)^{-\alpha} [(a_{22} - a_{11})(\omega_\alpha(t) - 1 - \omega_\alpha(t) \lambda 2^{-1}) f_1(t) \\ &\quad + a_{12}(2 - 2\lambda \omega_\alpha(t) - \lambda \omega_\alpha(x)) f_2(t)]\, dt, \end{aligned}$$

$$\begin{aligned} M_2[c_1, c_2, f_1(x), f_2(x)] &= \exp[\lambda \omega_\alpha(x)](b_{11} c_1 + \omega_\alpha c_2) + (b_{11}[(a_{22} - a_{11}) \omega_\alpha(x) + 2] \\ &\quad + b_{12}(x) 2^{-1}(a_{11} - a_{22})) f_1(x) + (b_{12}(x) a_{12} - 2a_{12} b_{11} \omega_\alpha(x)) f_2(x) \\ &- \int_a^x (t-a)^{-\alpha} \exp[\lambda(\omega_\alpha(x) - \omega_\alpha(t))] \{[b_{11}(\lambda(a_{22} - a_{11}) \omega_\alpha(t) + 2) - (a_{22} - a_{11}) \\ &+ b_{12}(x) \lambda 2^{-1}(a_{11} - a_{22})] f_1(t) + [2a_{12} - 2a_{12} \lambda \omega_\alpha(t) + b_{12}(x) a_{12}] f_2(t)\}\, dt, \end{aligned}$$

$$b_{12}(x) = b_{11}\omega_\alpha(x) + a_{12}^{-1}, \quad b_{11} = (2a_{12})^{-1}(a_{22} - a_{11}),$$

and c_1, c_2 are arbitrary constants.

Theorem 1.4. *Let in system (1.1) $\alpha > 1$, $n = 2$, the roots of characteristic equation (1.2) be real and equal, i.e. $\lambda_1 = \lambda_2 = \lambda > 0$. Let, moreover, $f_j(x)$, $a_{jk}(x) \in C(\overline{\Gamma})$ $(1 \leq j, k \leq 2)$ and, for $x \to a$, condition (1.6) and the following condition hold*

$$f_j(x) = o[(x-a)^\varepsilon], \quad \varepsilon > \alpha - 1, \quad j = 1, 2.$$

Then homogeneous system (1.1) has the trivial solution only. Inhomogeneous system (1.1) has a unique solution in the class $C(\overline{\Gamma})$:

$$y_j(x) = M_j[0,0,f_1(x),f_2(x)] + \sum_{m=1}^{2} \int_a^x \Gamma_{jm}(x,t) M_m[0,0,f_1(t),f_2(t)]\,dt,$$
$$j = 1, 2.$$

Here $\Gamma_{jm}(x,t)$ are the resolvents of the following system of integral Volterra equations with the weak singularity in the kernel

$$y_j(x) + \sum_{m=1}^{2} \int_a^x K_{jm}(x,t) y_m(t)\,dt = M_j[0,0,f_1(x),f_2(x)], \quad j = 1, 2,$$

and $K_{jm}(x,t)$ are the known functions dependent on the coefficients of system (1.1).

Theorem 1.5. *Let in system (1.1), $n = 2$, $\alpha > 1$, $f_j(x) \in C(\overline{\Gamma})$, $a_{jk}(x) = a_{jk} = \text{const}$. Suppose, moreover, that*

1) *the roots of characteristic equation (1.2) are complex conjugate, i.e., $\lambda_1 = p + iq$, $\lambda_2 = p - iq$, $p < 0$;*

2) *the functions $f_j(x)$, $(j = 1, 2)$ vanish in the neighborhood of the point $x = a$, and their behavior is as follows:*

$$f_j(x) = o[\exp(-|p|\omega_\alpha(x))(x-a)^{\gamma_7}], \quad \gamma_7 > \alpha - 1, \ j = 1, 2.$$

Then homogeneous system (1.1) has the two linearly independent solutions. Inhomogeneous system (1.1) is solvable in the class $C(\overline{\Gamma})$ and its solution is given by the formula

$$y_j(x) = E_j[c_1, c_2, f_1(x), f_2(x)], \quad j = 1, 2,$$

where c_1, c_2 *are arbitrary constants and*

$$E_1[c_1, c_2, f_1(x), f_2(x)] \equiv \exp[-|p|\omega_\alpha(x)][c_1\cos(q\omega_\alpha(x))] + c_2\sin(q\omega_\alpha(x))]$$
$$+ f_1(x) - \frac{a_{12}}{q}\int_a^x (t-a)^{-\alpha}\exp[p(\omega_\alpha(x)-\omega_\alpha(t))]\{[p\sin[q(\omega_\alpha(x)-\omega_\alpha(t))]$$
$$+ q\cos[q(\omega_\alpha(x)-\omega_\alpha(t))]f_2(t)] + [a_{11}q\cos[q(\omega_\alpha(t)-\omega_\alpha(x))]$$
$$+ (a_{11}a_{22}-a_{12}a_{21}-pa_{11})\sin[q(\omega_\alpha(t)-\omega_\alpha(x))]]f_1(t)\}\,dt,$$

$$E_2[c_1, c_2, f_1(x), f_2(x)] = \exp[-|p|\omega_\alpha(x)]\,\Big[\Big(\frac{p-a_{11}}{a_{12}}\Big)\cos(q\omega_\alpha(x))$$
$$-\frac{q}{a_{12}}\sin(q\omega_\alpha(x))c_1 + \Big(\frac{q}{a_{12}}\cos(q\omega_\alpha(x)) + \Big(\frac{p-a_{11}}{a_{12}}\Big)\sin(q\omega_\alpha(x))\Big)c_2\Big] + f_2(x)$$
$$-q^{-1}\int_a^x (t-a)^{-\alpha}\exp[p(\omega_\alpha(x)-\omega_\alpha(t))]\{[pa_{21}\sin[q(\omega_\alpha(x)-\omega)_\alpha(t))$$
$$+ a_{21}q\cos[q(\omega_\alpha(t)-\omega_\alpha(x))]]f_1(t) + [(2pq-qa_{11})\cos[q(\omega_\alpha(t)-\omega_\alpha(x))]$$
$$+ (q^2-p(p-a_{11}))\sin[q(\omega_\alpha(t)-\omega_\alpha(x))]f_2(t)\}\,dt.$$

Remark 1.1. The statement analogous to Theorem 1.5 is obtained in the case when $a_{jk}(x) \neq \text{const}$ and the condition of Theorem 1.5 hold. We suppose, moreover, that the functions $a_{jk}(x)$ satisfy the conditions similar to (1.6) in Theorem 1.3 (for $\lambda = p$). In this case the general solution of system (1.1) is expressed via the resolvents of the system of the integral Volterra equations of the second kind with the weak singularity in the point $t = a$ and with the right-hand side $E_j[c_1, c_2, f_1(x), f_2(x)]$, $j = 1, 2$.

Remark 1.2. The case when $\alpha = 1$, $n = 2$ and the roots of characteristic equation (1.2) are complex conjugate and $\operatorname{Re}\lambda = p < 0$ is studied also. The statement similar to Theorem 1.5 is obtained.

Remark 1.3. The analogous results hold in the case when $n = 2$, $\alpha \geq 1$ and the roots of characteristic equation (1.2) are complex conjugate and $\operatorname{Re}\lambda = p > 0$.

Theorems 1.1–1.5 and the statements mentioned in Remarks 1.1–1.3 are proved on the base of connection of system (1.1) and the system of linear ordinary differential equations with singular and super-singular coefficients [1] (see also [2, 3]).

2. THE GENERAL CASE

In this paper we research also the general linear system of Volterra equations

$$y_j(x) + \sum_{m=1}^{n} \int_a^x (t-a)^{-\alpha} K_{km}(x,t) y_m(t)\, dt = f_j(x), \quad 1 \le j \le n, \tag{2.1}$$

depending on the order of singularity α ($\alpha < 1$, $\alpha = 1$, $\alpha > 1$) and on the roots of correspondent characteristic equation

$$\Delta(\lambda) = \det \|K_{jm}(a,a) - \lambda \delta_{jm}\| = 0. \tag{2.2}$$

We represent system (2.1) as follows:

$$y_j(x) + \sum_{m=1}^{n} K_{jm}(a,a) \int_a^x (t-a)^{-\alpha} y_m(t)\, dt = F_j(x), \quad j = 1,2, \tag{2.3}$$

where

$$F_j(x) = f_j(x) - \sum_{m=1}^{n} \int_a^x (t-a)^{-\alpha} (K_{jm}(x,t) - K_{jm}(a,a)) y_m(t)\, dt$$
$$(1 \le j \le n). \tag{2.4}$$

Suppose that $\mu_j (1 \le j \le n)$ are the roots of characteristic equation (2.2)

$$\Delta(\mu) \equiv \det \|K_{jm}(a,a) - \mu \delta_{jm}\| = 0.$$

2.1. Let in system (2.1), $\alpha = 1$, $n = 2$ and the roots of characteristic equation (2.2) $\mu_j (j = 1,2)$ be real and distinct. Suppose, moreover, that $f_j(x) \in C(\overline{\Gamma})$, $K_{jm}(x,t) \in C(\overline{\mathbb{R}})$, $R = \{a < x < b,\ a < t < b\}$, $y_j(x) \in C(\overline{\Gamma})$ and the following condition holds for $x \to a$:

$$F_j(x) = o[(x-a)^{-(\lambda_1+\lambda_2)}], \quad j = 1,2. \tag{2.5}$$

If the solution of system (2.3) exists for $\mu_j < 0$ ($j = 1,2$) then, following Theorem 1.1, it is expressed as follows:

$$y_j(x) = K_{j,1}^{\mu_1,\mu_2}[c_1, c_2, F_1(x), F_2(x)], \quad j = 1,2, \tag{2.6}$$

where

$$K_{j,1}^{\mu_1,\mu_2}[c_1, c_2, F_1(x), F_2(x)] = \sum_{k=1}^{2} c_k c_{kj} (x-a)^{-\mu_k}$$
$$+ \Delta_0^{-1}[c_{1j}(c_{22}F_1(x) - c_{21}F_2(x)) + c_{2j}(c_{11}F_2(x) - c_{12}F_1)(x))] -$$

$$- \Delta_0^{-1} \int_a^x \Big[\Big(\frac{t-a}{x-a}\Big)^{\mu_1} c_{1j}\mu_1 (c_{22}F_1(t) - c_{21}F_2(t)) + \Big(\frac{t-a}{x-a}\Big)^{\mu_2} c_{2j}\mu_2 (c_{11}F_2(t) - c_{12}F_1(t)) \Big] \frac{dt}{t-a}, \quad (2.7)$$

where c_1, c_2 are arbitrary constants.

Now, substituting instead of the functions $F_j(x)$ $(j = 1, 2)$ their expressions from formula (2.4), we obtain the following system of integral Volterra equations

$$y_j(x) + \sum_{m=1}^{n} \int_a^x M_{jm}(x,t) y_m(t)\, dt = K_{j,1}^{\mu_1,\mu_2}[c_1, c_2, f_1(x), f_2(x)], \quad j = 1, 2, \quad (2.8)$$

where

$$M_{jm}(x,t) = \Delta_0^{-1}[c_{1j}\{c_{22}(K_{1m}(x,t) - K_{1m}(a,a)) - c_{21}(K_{2m}(x,t) - K_{2m}(a,a))\} + c_{2j}\{c_{11}(K_{2m}(x,t) - K_{1m}(a,a)) - c_{12}(K_{1m}(x,t) - K_{1m}(a,a))\}]/(t-a) - \frac{\Delta_0^{-1}}{t-a} \int_t^x \Big[\Big(\frac{s-a}{x-a}\Big)^{\mu_1} c_{1j}\mu_1 \{c_{22}(K_{1m}(s,t) - K_{1m}(a,a)) - c_{21}(K_{2m}(s,t) - K_{2m}(a,a))\} + \Big(\frac{s-a}{x-a}\Big)^{\mu_2} c_{2j}\mu_2 \{c_{11}(K_{2m}(s,t) - K_{1m}(a,a)) - c_{12}(K_{1m}(s,t) - K_{1m}(a,a))\}\Big] \frac{ds}{s-a}.$$

Assume that the functions $K_{jm}(x,t)$ satisfy the following conditions for $t \to x \to a$

$$K_{jm}(x,t) - K_{jm}(a,a) = o[(x-a)^{\beta_1}(t-a)^{\beta_2}], \quad (2.9)$$

$\beta_1 > 0$, $\beta_2 > |\mu_j| - 1$, $j = 1, 2$.

Then, system of integral equations (2.8) will be the system of integral Volterra equations of the second kind whose kernel will have the weak singularity in the point $t = a$.

If the kernels $K_{jm}(x,t)$ satisfy condition (2.9) for $x \to a$, $t \to a$, then system of integral equations (2.8) has a unique solution which is as follows:

$$y_j(x) = K_{j,1}^{\mu_1,\mu_2}[c_1, c_2, f_1(x), f_2(x)] - \sum_{m=1}^{2} \int_a^x \Gamma_{jm}(x,t) K_{m,1}^{\mu_1,\mu_2}[c_1, c_2, f_1(t), f_2(t)]\, dt, \quad j = 1, 2. \quad (2.10)$$

It is easy to see that the solution $y_j(x)$ of system (2.1) vanishes in the point $x = a$ and its behavior is defined by the asymptotic formula

$$y_j(x) = o[(x-a)^{|\mu_1|+|\mu_2|}], \quad j = 1, 2, \tag{2.11}$$

in the case when the functions $f_j(x)$ vanish in the point $x = a$ and

$$f_j(x) = o[(x-a)^{|\mu_1|+|\mu_2|+\delta}], \quad \delta > 0. \tag{2.12}$$

Thus, we have established the following theorem.

Theorem 2.1. *Suppose in system (2.3), $\alpha = 1$, $n = 2$, $f_j(x) \in C(\overline{\Gamma})$, $K_{jm}(x,t) \in C(\overline{\mathbb{R}})$ $(j, m = 1, 2)$ and the roots μ_j $(j = 1, 2)$ of characteristic equation (2.2) are negative and distinct. Let, moreover, the functions $f_j(x)$, $K_{jm}(x,t)$ satisfy conditions (2.12) and (2.9) when $x \to a$, $t \to a$. Then, homogeneous system (2.1) has two linearly independent solutions. Inhomogeneous system (2.1) is solvable in the class $C(\overline{\Gamma})$ and its solution contains two arbitrary constants. In this case, the general solution of system (2.1) is given by formula (2.10), where c_1, c_2 are arbitrary constants.*

In the case when $\alpha = 1$, $n = 2$, μ_j $(j = 1, 2)$ are real and distinct, $\mu_j > 0$, the following statement holds.

Theorem 2.2. *Suppose in system (2.1), $n = 2$, $\alpha = 1$ and the roots of characteristic equation (2.2) μ_j $(j = 1, 2)$ are real and distinct, $\mu_j > 0$. Let, moreover, $f_j(x) \in C(\overline{\Gamma})$, $K_{jm}(x,t) \in C(\overline{\mathbb{R}})$ $(j, m = 1, 2)$ and*

$$K_{jm}(x,t) - K_{jm}(a,a) = o((t-a)^{\varepsilon}), \quad \varepsilon > 0. \tag{2.13}$$

for $x \to a$, $t \to a$. Then inhomogeneous system (2.1) has the trivial solution only. Inhomogeneous system (2.1) has the unique solution which is given by formula (2.10) for $c_1 = c_2 = 0$.

In the case when $\alpha = 1$, $n = 2$, μ_j $(j = 1, 2)$ are real and distinct, $\mu_1 < 0$; $\mu_2 > 0$ the following statement holds.

Theorem 2.3. *Suppose in system (2.1) $\alpha = 1$, $n = 2$ and the roots of characteristic equation (2.2) are real and distinct and $\mu_1 < 0$, $\mu_2 > 0$. Let, moreover, $f_j(x) \in C(\overline{\Gamma})$, $K_{jm}(x,t) \in C(\overline{\mathbb{R}})$ $(j, m = 1, 2)$ and*

$$f_j(x) = o[(x-a)^{|\mu_2|}],$$
$$K_{jm}(x,t) - K_{jm}(a,a) = o[(x-a)^{\beta_1}], \qquad \beta_1 > |\mu_1| - 1$$

for $x \to a$, $t \to a$. Then homogeneous system (2.1) has the trivial solution only. Inhomogeneous system (2.1) is solvable in the class $C(\overline{\Gamma})$ and its solution has one arbitrary constants. The general solution of system (2.1) is given by the explicit formula via the resolvent of the following system of integral Volterra equations of the second kind with the weak singularity in the point $x = t = a$:

$$y_j(x) + \sum_{m=1}^{2} \int_a^x M_{jm}(x,t) y_m(t)\,\mathrm{d}t = K_{j,1}^{\mu_1,\mu_2}[c_1, 0, f_1(x), f_2(x)].$$

Remark 2.1. Suppose in system of equations (2.1), $n = 2$, $\alpha = 1$, the roots of characteristic equation (2.2) μ_1 and μ_2 are real, distinct and $\mu_1 > 0$, $\mu_2 < 0$. Let, moreover, $f_j(x) \in C(\overline{\Gamma})$, $K_{jm}(x,t) \in C(\overline{\mathbb{R}})$ and satisfy the conditions

$$f_j(x) = o[(x-a)^{|\mu_2|}],$$
$$K_{jm}(x,t) - K_{jm}(a,a) = o[(x-a)^{\beta_2}], \qquad \beta_2 > |\mu_2| - 1$$

for $x \to a$, $t \to a$. Then the statement similar to Theorem 2.2 holds for system (2.1).

2.2. Let in system (2.1) $n = 2$, $\alpha > 1$, the roots of characteristic equation (2.2) μ_j $(j = 1, 2)$ be real, distinct and $\mu_j < 0$. Let $f_j(x) \in C(\overline{\Gamma})$, the functions $K_{jm}(x,t)$ and the solutions $y_j(x)$, $j, m = 1, 2$ of system (2.1) be such that $F_j(x) \in C(\overline{\Gamma})$ and

$$F_j(x) = o[\exp[(\mu_1 + \mu_2)\omega_\alpha(x)](x-a)^{\gamma_\varepsilon}], \;\; \gamma_\varepsilon > \alpha - 1, \; j = 1, 2. \tag{2.14}$$

for $x \to a$. In this case, when $\alpha > 1$, using the results of Theorem 1.1 it is easy to prove that the general solution of equation (2.3) is given in Γ by the formula

$$y_j(x) = K_{j,\alpha}^{\mu_1,\mu_2}[c_1, c_2, F_1(x), F_2(x)], \quad j = 1, 2, \tag{2.15}$$

where

$$\begin{aligned} K_{j,\alpha}^{\mu_1,\mu_2}[c_1, c_2, F_1(x), F_2(x)] &\equiv \sum_{m=1}^{2} c_m \exp(\nu_m \omega_\alpha(x)) \\ &+ \Delta_0^{-1}[c_{1j}(c_{22}F_1(x) - c_{21}F_2(x)) + c_{2j}(c_{11}F_2(x) - c_{12}F_1(x))] \\ &- \Delta_0^{-1} \int_a^x \{\exp[\lambda_1(\omega_\alpha(x) - \omega_\alpha(t))]\lambda_1 c_{1j}(F_1(t)c_{22} - F_2(t)c_{21}) \\ &+ \exp[\lambda_2(\omega_\alpha(x) - \omega_\alpha(t))]\lambda_2 c_{2j}(c_{11}F_2(t) - c_{12}F_1(t))](t-a)^{-\alpha}\,\mathrm{d}t, \\ &\qquad j = 1, 2, \end{aligned} \tag{2.16}$$

c_1, c_2 are arbitrary constants.

In representation (2.15) substituting instead of $F_j(x)$, $(j = 1, 2)$ their expressions from formula (2.4), we obtain the following system of Volterra equations for unknown functions $y_j(x)$:

$$y_j(x) + \sum_{m=1}^{n} \int_a^s N_{jk}(x,t) y_k(t)\,\mathrm{dt} = K_{j,\alpha}^{\mu_1,\mu_2}[c_1, c_2, f_1(x), f_2(x)], \quad j = 1, 2, \tag{2.17}$$

where

$$\begin{aligned}
N_{jm} &= \Delta_0^{-1}[c_{1j}\{c_{22}(K_{1m}(x,t) - K_{1m}(a,a)) - c_{21}(K_{2m}(x,t) - K_{2m}(a,a))\} \\
&\quad + c_{2j}(K_{2m}(x,t) - K_{2m}(a,a)) - c_{12}(K_{1m}(x,t) - K_{1m}(a,a))\}](t-a)^{-\alpha} \\
&\quad - \Delta_0^{-1}(t-a)^{-\alpha} \int_t^x [\exp[\mu_1(\omega_\alpha(x) - \omega_\alpha(s))] c_{1j}\mu_1\{c_{22}(K_{1m}(x,t) \\
&\quad - K_{1m}(a,a)) - c_{21}(K_{2m}(s,t) - K_{2m}(a,a))\} + \exp[\mu_2(\omega_\alpha(x) - \omega_\alpha(s))] \\
&\quad \times c_{2j}\mu_2\{c_{11}(K_{2m}(s,t) - K_{2m}(a,a)) - c_{12}(K_{1m}(s,t) - K_{1m}(a,a))\}](s-a)^{-\alpha}\,\mathrm{ds}.
\end{aligned}$$

Suppose the functions $f_j(x)$ satisfy the following conditions for $x \to a$:

$$\begin{gathered}
f_j(x) = o[\exp((\mu_1 + \mu_2)\omega_\alpha(x))(x-a)^{\gamma_9}], \\
\mu_j < 0, \quad \gamma_9 > \alpha - 1, \quad j = 1, 2.
\end{gathered} \tag{2.18}$$

Then the functions

$$\Omega_j(x) = K_{j,\alpha}^{\mu_1,\mu_2}[c_1, c_2, f_1(x), f_2(x)], \quad j = 1, 2$$

will be continuous, i.e. $\Omega_j(x) \in C(\overline{\Gamma})$ and their asymptotic behaviour for $x \to a$ is as follows:

$$\Omega_j(x) = o[\exp(-\mu\omega_\alpha(x))], \quad \mu = \min(|\mu_1|, |\mu_2|).$$

Then the kernels $N_{jm}(x,t)$ of system (2.17) will have the weak singularity if the functions $K_{jm}(x,t)$ satisfy the conditions

$$\begin{gathered}
K_{jm}(x,t) - K_{jm}(a,a) = o\Big[\exp\Big(\sum_{m=1}^{2} \mu_m \omega_\alpha(t)\Big)(x-a)^{\gamma_{10}}\Big], \\
\gamma_{10} > \alpha - 1, \quad j = 1, 2
\end{gathered} \tag{2.19}$$

in the neighborhood of the points $t = a$, $x = a$.

We see that if conditions (2.18), (2.19) hold and $y_j(x) \in C(\overline{\Gamma})$, then the functions $F_j(x)$, $(j = 1, 2)$ satisfy conditions (2.14).

Thus, in the case when the roots of characteristic equation (2.2) are real and distinct, $n = 2$, $\alpha > 1$, the following statement holds.

Theorem 2.4. *Suppose in system (2.1), $n = 2$, $\alpha > 1$, the roots μ_j ($j = 1, 2$) of characteristic equation (2.2) are real and distinct, and $\mu_j < 0$. Let, moreover, (2.19) hold for $x \to a$, $t \to a$. Then, homogeneous system (2.1) has two linearly independent solutions. Inhomogeneous system (2.1) is solvable in the class $C(\overline{\Gamma})$ and its general solution has two arbitrary constants. The general solution of system (2.1) is represented in the form*

$$y_j(x) = K_{j,\alpha}^{\mu_1,\mu_2}[c_1, c_2, f_1(x), f_2(x)] - \sum_{m=1}^{2} \int_a^x \Gamma_{jm}(x,t) K_{m,\alpha}^{\mu_1,\mu_2}[c_1, c_2, f_1(t), f_2(t)]\, dt, \quad j = 1, 2, \quad (2.20)$$

where $j = 1, 2$; $\Gamma_{jm}(x, t)$ are the resolvent of the system of integral Volterra equations (2.17) with the weak singularity; c_1, c_2 are arbitrary constants.

In the case when $\alpha > 1$, $n = 2$, μ_j ($j = 1, 2$) are positive and distinct, the following statement holds.

Theorem 2.5. *Suppose in system (2.1), $\alpha > 1$, $n = 2$ and the roots μ_j ($j = 1, 2$) of characteristic equation (2.2) are real, distinct and positive. Let, moreover, $f_j(x) \in C(\overline{\Gamma})$, $K_{jm}(x, t) \in C(\overline{\mathbb{R}})$ ($m = 1, 2$) and satisfy the condition*

$$K_{jm}(xmt) - K_{jm}(a, a) = o[(t - a)^{\gamma_{11}}], \quad \gamma_{11} > \alpha - 1$$

for $t \to a$. Then, homogeneous system (2.1) has the trivial solution only. Inhomogeneous system (2.1) has a unique solution which is given by formula (2.20) for $c_1 = c_2 = 0$.

Remark 2.2. The statement similar to Theorem 2.4 and Remark 2.1 is obtained in the cases when $\alpha > 1$, $n = 2$, the roots μ_j ($j = 1, 2$) of the characteristic equation (2.2) are real, distinct and $\mu_1 < 0$, $\mu_2 > 0$.

2.3. Suppose in system (2.1) $\alpha > 1$, $n = 2$ and the roots μ_j ($j = 1, 2$) of characteristic equation (2.2) are real and multiple. Let, moreover, $f_j(x) \in C(\overline{\Gamma})$, $K_{jm}(x, t) \in C(\overline{\mathbb{R}})$, $y_j(x) \in C(\overline{\Gamma})$ are such that the functions $F_j(x)$ satisfy the conditions

$$F_j(x) = o[\exp(-|\mu|\omega_\alpha(x))(x - a)^{\gamma_{11}}], \quad \gamma_{11} > 2(\alpha - 1),$$

for $x \to a$ and $\mu_1 = \mu_2 = \mu < 0$. Then, if there exists the solution of system (2.3) when $\mu < 0$, then it is expressed in the form (Corollary 1.3)

$$y_j(x) = M_j[c_1, c_2, F_1(x), F_2(x)], \quad j = 1, 2. \tag{2.21}$$

In representation (2.21), instead of $F_1(x)$, $F_2(x)$, we substitute their expressions from formula (2.4). As a result we obtain the following system of integral Volterra equations

$$y_j(x) + \sum_{m=1}^{2} \int_a^x T_{jm}(x,t) y_m(t)\, dt = M_j[c_1, c_2, f_1(x), f_2(x)], \quad j = 1, 2, \tag{2.22}$$

where $T_{jm}(x,t)$ are set. If, moreover, the functions $K_{jm}(x,t)$ satisfy the following conditions in the neighborhood of the points $x = a$, $t = a$

$$K_{jm}(x,t) - K_{jm}(a,a) = o[\exp(\lambda\omega_\alpha(x))(x-a)^{\gamma_{12}}(t-a)^{\gamma_{13}}],$$
$$\gamma_{12} > \alpha - 1, \quad \gamma_{13} > \alpha - 1, \quad j, m = 1, 2, \tag{2.23}$$

then $T_{jm}(x,t)$ has the weak singularity in the point $t = a$ and is continuous when $x = a$.

Thus, we have proved the following theorem.

Theorem 2.6. *Suppose in system* (2.1), $\alpha > 1$, $n = 2$, $f_j(x) \in C(\overline{\Gamma})$, $K_{jm}(x,t) \in C(\overline{\mathbb{R}})$ $(j, m = 1, 2)$ *and the roots of characteristic equation* (2.2) *are real and equal:* $\mu_1 = \mu_2 = \mu < 0$. *Let, moreover,* $f_j(x)$, $K_{jm}(x,t)$ *satisfy conditions* (1.5), (2) *for* $x \to a$, $t \to a$. *Then homogeneous system* (2.1) *has two linearly independent solutions. Inhomogeneous system* (2.1) *is solvable in the class* $C(\overline{\Gamma})$ *and its general solution has two arbitrary constants. In this case the general solution of system* (2.1) *is represented explicitly via the resolvents of system* (2.22).

Remark 2.3. Suppose in system (2.1), $\alpha > 1$, $n = 2$, the roots of characteristic equation (2.2) μ_j $(j = 1, 2)$ are real and distinct and have distinct signs. Then the statement similar to Theorem 2.4 and Remark 2.1 holds for system (2.1).

Remark 2.4. Let in system (2.1), $\alpha = 1$, $n = 2$, the roots of characteristic equation (2.2) be real and multiple: either $\mu_1 = \mu_2 = \mu < 0$ or $\mu_1 = \mu_2 = \mu > 0$, or $\mu_1 = \mu_2 = 0$. Then the statement holds similar to Theorems 2.1, 2.2.

Remark 2.5. System (2.1) is stated also in the case $n = 2$, $\alpha \geq 1$, $\mu_1 = p + iq$, $\mu_2 = p - iq$, $p > 0$, $p < 0$ or $p = 0$. In this case the statement similar to Theorems 2.1–2.5 for solvability of system (2.1) holds.

Remark 2.6. Depending on the roots of characteristic equation (2.2), system (2.1) is stated in the cases $n > 2$, $\alpha < 1$; $n > 2$, $\alpha = 1$; $n > 2$, $\alpha > 1$. In these cases the statement similar to Theorems 1.1, 1.2 hold also.

Remark 2.7. The linear system of Volterra integral equations with the fixed singular or super-singular point $x = b$ is investigated. This is the system

$$y_j(x) + \sum_{m=1}^{n} \int_x^b (b-t)^{-\beta} K_{jm}(x,t) y_m(t)\, \mathrm{d}t = f_j(x), \quad 1 \leq j \leq m,$$

with various order of singularity β ($\beta < 1$; $\beta = 1$; $\beta > 1$). Depending on the order of singularity and the roots of the characteristic equation

$$\Delta(\lambda) = \det \|K_{jm}(b,b) + \lambda \delta_{jm}\| = 0$$

the statement similar to theorems above hold.

REFERENCES

1. N. Rajabov, *Introduction to Ordinary Differential Equations with Singular and Super-Singular Coefficients.* Dushanbe, 1998.

2. N. Rajabov, Higher order ordinary differential equations with super-singular points. In: *Partial Differential and Integral Equations.* Kluwer Academic Publisher, 1999, 348–358.

3. N. Rajabov, Explicit solution of certain classes of linear integral Volterra equations with left and right fixed singular and super-singular points in the kernel. In: *Problems of Mathematics and Informatics.* Dushanbe, 2001, 31–34 (in Russian).

4. N. Rajabov, Systems of linear integral equations of Volterra type with singular and super-singular kernels. In: *Ill-Posed and Nonclassical Problems of Mathematical Physics and Analysis.* Samarkand, Uzbekistan, September 11–15, 2000, p. 73.

5. N. Rajabov, To the theory of one class of systems of integral Volterra equations with an immovable singular and super-singular point. *Proc.*

of Intern. Sci. Conf. Methods of Functional Theory and their Applications. Dushanbe, 2000, 33–34 (in Russian).

6. N. Rajabov, General integral equations of Volterra type with the left and right immovable singular and super-singular points in the kernel. *Izv. Akad. Nauk. Tadj. Otdel Fiz. Mat. Chemical and Geolog. Sci.* (2001) **1**, 30–46 (in Russian)

Ill-Posed and Non-Classical Problems of
Mathematical Physics and Analysis, Samarkand, 2000, pp. 125–141
M. M. Lavrent'ev and S. I. Kabanikhin (Eds)

Inverses of a family of bounded linear operators, generalized pythagorean theorems and reproducing kernels

S. SAITOH*

Abstract — When we consider the inversion of a linear mapping in the framework of Hilbert spaces, we need naturally the theory of reproducing kernels. We considered recently generalizations of the Pythagorean theorem with geometric meanings and from the generalizations we can obtain a general and fundamental concept for the inversion of a family of bounded linear operators on a Hilbert space into various Hilbert spaces with any index set. After reviewing the applications of the general theory of reproducing kernels, we shall state the results for the case of operator versions. As an announcement of our recent results as typical examples, we shall state explicit inversion formulas of a family of matrices and explicit generalizations of the Pythagorean theorem. We shall refer to solutions and generalized solutions of general bounded linear operator equations with continuous parameters on a Hilbert space.

1. REPRODUCING KERNELS

We consider any positive matrix $K(p,q)$ on E; that is, for an abstract set E and for a complex-valued function $K(p,q)$ on $E \times E$, it satisfies that for any finite points $\{p_j\}$ of E and for any complex numbers $\{C_j\}$,

$$\sum_j \sum_{j'} C_j \overline{C_{j'}} K(p_{j'}, p_j) \geq 0.$$

Then, by the fundamental theorem by Moore—Aronszajn, we have:

*Department of Mathematics, Faculty of Engineering, Gunma University, Kiryu 376-8515, Japan. E-mail: ssaitoh@math.sci.gunma-u.ac.jp

Proposition 1.1 [1]. For any positive matrix $K(p,q)$ on E, there exists a uniquely determined functional Hilbert space H_K (RKHS H_K) comprising complex-valued functions $\{f\}$ on E and admitting the reproducing kernel $K(p,q)$ satisfying and characterized by

$$K(\cdot,q) \in H_K \qquad \text{for any} \quad q \in E \tag{1.1}$$

and, for any $q \in E$ and for any $f \in H_K$

$$f(q) = (f(\cdot), K(\cdot,q))_{H_K}. \tag{1.2}$$

For some general properties for reproducing kernel Hilbert spaces and for various constructions of the RKHS H_K from a positive matrix $K(p,q)$, see the recent book [15] and its Chapter 2, Section 5, respectively.

2. CONNECTIONS WITH LINEAR MAPPINGS

Let us connect linear mappings in the framework of Hilbert spaces with reproducing kernels [8].

For an abstract set E and for any Hilbert (possibly finite-dimensional) space H, we shall consider an H-valued function $\boldsymbol{h}$ on E

$$\boldsymbol{h}: \quad E \longrightarrow H \tag{2.1}$$

and the linear mapping for H

$$f(p) = (\boldsymbol{f}, \boldsymbol{h}(p))_H \quad \text{for} \quad \boldsymbol{f} \in H \tag{2.2}$$

into a linear space comprising complex-valued functions on E. For this linear mapping (2.2), we form the positive matrix $K(p,q)$ on E defined by

$$K(p,q) = (\boldsymbol{h}(q), \boldsymbol{h}(p))_H \quad \text{on} \quad E \times E. \tag{2.3}$$

Then, we have the following fundamental results:

(I) For the RKHS H_K admitting the reproducing kernel $K(p,q)$ defined by (2.3), the images $\{f(p)\}$ by (2.2) for H are characterized as the members of the RKHS H_K.

(II) In general, we have the inequality in (2.2)

$$\|f\|_{H_K} \leq \|\boldsymbol{f}\|_H, \tag{2.4}$$

however, for any $f \in H_K$ there exists a uniquely determined $\boldsymbol{f}^* \in H$ satisfying

$$f(p) = (\boldsymbol{f}^*, \boldsymbol{h}(p))_H \qquad \text{on} \quad E \tag{2.5}$$

and

$$\|f\|_{H_K} = \|\boldsymbol{f}^*\|_H. \tag{2.6}$$

In (2.4), the isometry holds if and only if $\{\boldsymbol{h}(p); p \in E\}$ is complete in H.

(III) We can obtain the inversion formula for (2.2) in the form

$$f \longrightarrow \boldsymbol{f}^*, \tag{2.7}$$

by using the RKHS H_K.

However, this inversion formula will depend on, case by case, the realizations of the RKHS H_K.

(IV) Conversely, if we have an isometric mapping $\tilde{L}$ from a RKHS H_K admitting a reproducing kernel $K(p,q)$ on E onto a Hilbert space H, then the mapping is linear and its isometric inversion $\tilde{L}^{-1}$ is represented in the form (2.2). Here, the Hilbert space H-valued function $\boldsymbol{h}$ satisfying (2.1) and (2.2) is given by

$$\boldsymbol{h}(p) = \tilde{L}K(\cdot\,, p) \qquad \text{on} \quad E \tag{2.8}$$

and, then $\{\boldsymbol{h}(p); p \in E\}$ is complete in H.

When (2.2) is isometrical, sometimes we can use the isometric mapping for a realization of the RKHS H_K, conversely — that is, if the inverse L^{-1} of the linear mapping (2.2) is known, then we have $\|f\|_{H_K} = \|L^{-1}f\|_H$.

We shall state some general applications of the results (I)–(IV) to several wide subjects and their basic references:

(1) Linear mappings [8, 12]. The fact that the image space of a linear mapping in the framework of Hilbert spaces is characterized as the reproducing kernel Hilbert space defined by (2.3) is the most important one in the general theory of reproducing kernels. Therefore, the fact will mean that the theory of reproducing kernels is fundamental and a general concept in mathematics, as in the idea of Hilbert spaces. To look for the characterization of the image space is a starting point when we consider the linear equation (2.2). (II) gives a generalization of the Pythagorean theorem (see also [7]) and means that in the general linear mapping (2.2) there exists essentially an isometric identity between the input and the output. (III) gives a generalized (natural) inverse (solution) of the linear mapping (equation) (2.2). (IV) gives a general method determining and constructing the linear

system from an isometric relation between outputs and inputs by using the reproducing kernel in the output space.

In particular, when we write (2.2) in an integral transform in the framework of Hilbert spaces, it becomes a Fredholm integral equation of the first kind and it was thought that to solve the equation is a typical ill-posed problem. At this point, we can see why we meet the ill-posed problem; that is, we do not consider it in the natural image space H_K, but in some artificial spaces. Our method will give a general method solving the Fredholm integral equations of the first kind in the framework of Hilbert spaces. We gave new solutions of the many typical integral equations like the Laplace transform, the Weierstrass transform, the Cauchy integral transform and the Poisson integral equation. See [12] and [15].

(2) Linear mappings among smooth functions [19]. We considered linear mappings in the framework of Hilbert spaces, however, we can also consider linear mappings in the framework of reproducing kernel Hilbert spaces comprising smooth functions, similarly. Conversely, reproducing kernel Hilbert spaces are considered as the images of some Hilbert spaces by considering some Hermitian representations (2.3) of the reproducing kernels. Such decompositions of reproducing kernels are, in general, possible. This idea is important in [19] and also in the following items (6) and (7).

(3) Nonharmonic linear mappings [9]. If the linear system vectors $\boldsymbol{h}(p)$ move in a small way (perturbation of the linear system) in the Hilbert space H, then we can not calculate the related positive matrix (2.3), however, we can discuss the inversion formula and an isometric identity of the linear mapping, by restricting the image functions in a subspace. The prototype result is the Paley—Wiener theorem on nonharmonic Fourier series.

(4) Various norm inequalities [9, 13]. Relations among positive matrices correspond to those of the associated reproducing kernel Hilbert spaces, by the minimum principle. So, we can derive various norm inequalities among reproducing kernel Hilbert spaces. We were able to derive many beautiful norm inequalities.

(5) Nonlinear mappings [13, 16]. In a very general nonlinear mapping of a reproducing kernel Hilbert space, we can look for a natural reproducing kernel Hilbert space containing the image space and furthermore, we can derive a natural norm inequality in the nonlinear mapping. How to catch nonlinearity in connection with linearity? It seems that the theory of reproducing kernels gives a fundamental and interesting answer for this basic question.

(6) Linear integral equations [20].

(7) Linear differential equations with variable coefficients [20]. In linear integro-differential equations with general variable coefficients, we can discuss the existence and construction of the solutions, if the solutions exist. This method is called a *backward transformation method* and by reducing the equations to Fredholm integral equations of the first kind—(2.2)—and we can discuss the classical solutions, in general linear equations.

(8) Approximation theory [3, 2]. Reproducing kernel Hilbert spaces are very nice function spaces, because the point evaluations are continuous. Then, the reproducing kernels are a fundamental tool in the related approximation theory.

(9) Representations of inverse functions [14]. For an arbitrary mapping from an abstract set into an abstract set, we discussed the problem of representing its inverse in term of the direct mapping and we derived a unified method for this problem. As a simple example, we can represent the Taylor coefficients of the inverse of the Riemann mapping function on the unit disc on the complex plane in terms of the Riemann mapping function. This fact was important in the representation of analytic functions in terms of local data in [23, 24].

(10) Various operators among Hilbert spaces [17]. Among various abstract Hilbert spaces, we can introduce various operators of sum, product, integral and derivative by using the linear mapping (2.2) or very general nonlinear mappings. The prototype operator is convolution and we discussed it from a wide and general viewpoint with concrete examples. See [25, 26], and [27] for further developments in connection with applications to stability in inverse problems.

(11) Sampling theorems [15, Chapter 4, Section 2; 5]. The Whittaker—Kotel'nikov—Shannon sampling theorem may be interpretated by (I) and (II), very well and we can discuss the truncation error estimates in the sampling theory. Higgins [5] established a fully general theory for [15].

(12) Interpolation problems of Pick—Nevanlinna type [9, 10]. A general and abstract theory of Pick—Nevanlinna interpolation problems may be discussed by using the general theory of reproducing kernels.

(13) Analytic extension formulas and their applications [21, 11]. We were able to obtain various analytic extension formulas and their applications from various isometric identities (II). For their applications to nonlinear partial differential equations, see the survey article by Hayashi [4].

In this survey article, we shall present also new results on

(14) Inversions of a family of bounded linear operators on a Hilbert space into various Hilbert spaces, which are generalizations of [22] and [7].

As an announcement of our recent results as typical applications we shall state explicit inversion formulas of a family of matrices and explicit generalizations of the Pythagorean theorem for right triangles. Furthermore, in connection with Kaczmarz's Method for a finite number of bounded linear operator equations on a Hilbert space, we shall give our generalized solutions for general operator equations with continuous parameters.

3. OPERATOR VERSIONS

We shall give operator versions of the fundamental theory (I)–(IV) which may be expected to have many concrete applications. In particular, for full generalizations of the Pythagorean theorem with geometric meanings, see [7]. Some special versions were given in [22]. We will give more direct and clear results in matrix theory in Section 5.

For an abstract set $\mathbf{\Lambda}$, we shall consider an operator-valued function L_λ on $\mathbf{\Lambda}$,

$$\mathbf{\Lambda} \longrightarrow L_\lambda \tag{3.1}$$

where L_λ are bounded linear operators from a Hilbert space H into various Hilbert spaces $\boldsymbol{H}_\lambda$,

$$L_\lambda : H \longrightarrow \boldsymbol{H}_\lambda. \tag{3.2}$$

In particular, we are interested in the inversion formula

$$L_\lambda x \longrightarrow x, \quad x \in H. \tag{3.3}$$

Here, we consider $\{L_\lambda x; \lambda \in \Lambda\}$ as informations obtained from x and we wish to determine x from the informations. Indeed, for one object x, we will, in general, obtain many type informations $L_\lambda x$. However, the informations $L_\lambda x$ belong to various Hilbert spaces $\boldsymbol{H}_\lambda$, and so, in order to unify the informations in a sense, we shall take fixed elements $\boldsymbol{b}_{\lambda,\omega} \in \boldsymbol{H}_\lambda$ and consider the linear mapping from H

$$X_{\boldsymbol{b}}(\lambda,\omega) = (L_\lambda x, \boldsymbol{b}_{\lambda,\omega})_{\boldsymbol{H}_\lambda} = (x, L_\lambda^* \boldsymbol{b}_{\lambda,\omega})_H, \quad x \in H$$

into a linear space comprising functions on $\mathbf{\Lambda} \times \mathbf{\Omega}$. For the informations $L_\lambda x$, we shall consider $X_{\boldsymbol{b}}(\lambda,\omega)$ as observations (measurements, in fact) for x depending on λ and ω. For this linear mapping (3.4), we form the positive matrix $K_{\boldsymbol{b}}(\lambda, \omega; \lambda', \omega')$ on $\mathbf{\Lambda} \times \mathbf{\Omega}$ defined by

$$\begin{aligned} K_{\boldsymbol{b}}(\lambda, \omega; \lambda', \omega') &= (L^*_{\lambda'} \boldsymbol{b}_{\lambda',\omega'}, L^*_{\lambda} \boldsymbol{b}_{\lambda,\omega})_H \\ &= (L_{\lambda} L^*_{\lambda'} \boldsymbol{b}_{\lambda',\omega'}, \boldsymbol{b}_{\lambda,\omega})_{\boldsymbol{H}_{\lambda}} \quad \text{on } \mathbf{\Lambda} \times \mathbf{\Omega}. \end{aligned} \tag{3.4}$$

Then, as in (I)–(IV), we have the following fundamental results:

(I′) For the RKHS $H_{K_{\boldsymbol{b}}}$ admitting the reproducing kernel $K_{\boldsymbol{b}}(\lambda, \omega; \lambda', \omega')$ defined by (3.4), the images $\{X_{\boldsymbol{b}}(\lambda, \omega)\}$ by (3.4) for H are characterized as the members of the RKHS $H_{K_{\boldsymbol{b}}}$.

(II′) In general, we have the inequality in (3.4)

$$\|X_{\boldsymbol{b}}\|_{H_{K_{\boldsymbol{b}}}} \leq \|x\|_H, \tag{3.5}$$

however, for any $X_{\boldsymbol{b}} \in H_{K_{\boldsymbol{b}}}$ there exists a uniquely determined $x' \in H$ satisfying

$$X_{\boldsymbol{b}}(\lambda, \omega) = (x', L^*_{\lambda} \boldsymbol{b}_{\lambda,\omega})_H \quad \text{on} \quad \mathbf{\Lambda} \times \mathbf{\Omega} \tag{3.6}$$

and

$$\|X_{\boldsymbol{b}}\|_{H_{K_{\boldsymbol{b}}}} = \|x'\|_H. \tag{3.7}$$

In (3.5), the isometry holds if and only if $\{L^*_{\lambda} \boldsymbol{b}_{\lambda,\omega} \mid (\lambda, \omega) \in \mathbf{\Lambda} \times \mathbf{\Omega}\}$ is complete in H.

(III′) We can obtain the inversion formula for (3.4) and so, for the mapping (3.3) as in (III), in the form

$$L_{\lambda} x \longrightarrow (L_{\lambda} x, \boldsymbol{b}_{\lambda,\omega})_{\boldsymbol{H}_{\lambda}} = X_{\boldsymbol{b}}(\lambda, \omega) \longrightarrow x', \tag{3.8}$$

by using the RKHS $H_{K_{\boldsymbol{b}}}$.

(IV′) Conversely, if we have an isometric mapping $\tilde{L}$ from a RKHS $H_{K_{\boldsymbol{b}}}$ admitting a reproducing kernel $K_{\boldsymbol{b}}(\lambda, \omega; \lambda', \omega')$ on $\mathbf{\Lambda} \times \mathbf{\Omega}$ in the form (3.5) using bounded linear operators L_{λ} and fixed vectors $\mathbf{b}_{\lambda,\omega}$ onto a Hilbert space H, then the mapping $\tilde{L}$ is linear and the isometric inversion $\tilde{L}^{-1}$ is represented in the form (3.4) by using

$$L^*_{\lambda} \boldsymbol{b}_{\lambda,\omega} = \tilde{L} K_{\boldsymbol{b}}(\cdot, \cdot; \lambda, \omega) \quad \text{on } \mathbf{\Lambda} \times \mathbf{\Omega}. \tag{3.9}$$

Further, then $\{L^*_{\lambda} \boldsymbol{b}_{\lambda,\omega} \mid (\lambda, \omega) \in \mathbf{\Lambda} \times \mathbf{\Omega}\}$ is complete in H.

The author obtained the above concept for the operator versions from a generalization of the Pythagorean theorem in the following way:

Let $\boldsymbol{x} \in \mathbb{R}^n$ and $\{\boldsymbol{e}_j\}_{j=1}^n$ be linearly independent unit vectors. We consider the linear mappings

$$L : \boldsymbol{x} \longrightarrow \{\boldsymbol{x} - (\boldsymbol{x}, \boldsymbol{e}_j)\boldsymbol{e}_j\}_{j=1}^n \tag{3.10}$$

from $\mathbb{R}^n$ into $\mathbb{R}^n$. Then we wish to establish an isometric identity and inversion formula in the operators. Recall the Pythagorean theorem for $n = 2$. By our operator versions, we can establish the desired results.

Note that in (3.11), for $n \geq 3$ if we consider

$$\{\|\boldsymbol{x} - (\boldsymbol{x}, \boldsymbol{e}_j)\boldsymbol{e}_j\|\}_{j=1}^n \tag{3.11}$$

as scalar valued mappings, then the mappings are not linear more. So, we must consider the operator valued mappings in the problems.

We see that some related equations were considered as in the following way [6, pp. 128–157]:

Let H, H_j, $j = 1, 2, \ldots, p$ be Hilbert spaces and let

$$R_j : H \longrightarrow H_j, \quad j = 1, 2, \ldots, p \tag{3.12}$$

be linear continuous maps from H onto H_j. Let $g_j \in H_j$ be given. Then, consider the problem to compute $f \in H$ such that

$$R_j f = g_j, \quad j = 1, 2, \ldots, p. \tag{3.13}$$

This equations are very important in the theory of computerized tomography by the discretization. The typical method is Kaczmarz's Method based on an iterative method by using the orthogonal projections P_j in H onto the affine subspaces $R_j f = g_j$.

Our direct solutions for (3.14) seem that the result is stable for the sake of the use (3.14) as data, because we use (3.9) which is given by the inner product.

In general, in equations (3.14) we have noises and errors for the data g_j and so, in those cases the equations do, in general, not have solutions. So, we will consider a more general solution which is called a generalized solution (inverse) in Section 6.

4. INVERSES OF A FAMILY OF MATRICES

If a reproducing kernel Hilbert space is finite dimensional, then the theory of reproducing kernels is reduced to that of matrices [15]. Furthermore, the results derived from the theory of reproducing kernels will be derived also directly by that of matrices. In those two sections, we shall state the results in the matrix theory derived from the above general concept explicitly and in the framework of matrix theory, for its importance. Of course, the results obtained here are more sharp and more concrete than those in [3]. Indeed, from the theory of reproducing kernels, we can not discuss the linearly independence of the matrix Q_b in Section 5, which is introduced by using an auxiliary vector $\boldsymbol{b}$.

A concept and a basic interest in this section are as follows:

For any matrix A of type $m \times n$, we consider the linear mapping

$$A\boldsymbol{x} = \boldsymbol{y} \tag{4.1}$$

from $\mathbb{R}^n$ into $\mathbb{R}^m$. Then its inversion is of course examined in detail. However, in many cases, we shall consider a general mapping

$$A_k^{j_k}\boldsymbol{x} = \boldsymbol{y}_k^{j_k} \tag{4.2}$$

from $\mathbb{R}^n$ into $\mathbb{R}^{j_k}$ for various j_k and we consider its inversion, because we have, in general, many type data $\boldsymbol{y}_k^{j_k}$ for $\boldsymbol{x}$. Here, the important fact is that each data $\boldsymbol{y}_k^{j_k}$ for $\boldsymbol{x}$ is one unit and it is not decomposed into components, otherwise, the problem (4.2) will be reduced to the case (4.1) by considering the augmented matrix

$$\begin{pmatrix} A_1^{j_1} \\ A_2^{j_2} \\ \vdots \\ A_s^{j_s} \end{pmatrix}. \tag{4.3}$$

Furthermore, we shall measure the data $\boldsymbol{y}_k^{j_k}$ as the scalar value

$$\langle \boldsymbol{y}_k^{j_k}, \boldsymbol{b}_{j_k} \rangle_{\mathbb{R}^{j_k}}$$

in $\mathbb{R}^{j_k}$ by taking a vector $\boldsymbol{b}_{j_k}$ of $\mathbb{R}^{j_k}$. This concept comes from a general approach for inversion formulas for a family of bounded linear operators on a Hilbert space using the theory of reproducing kernels in Section 3. We shall give concrete examples in our full general Pythagorean theorem.

In order to simplify the representation of the results, we shall consider all on the real field $\mathbb{R}$.

We see the reference [6, pp. 128–157] for the equation (4.2) and the equation was examined by many authors in connection with Computerized Tomography. One typical method to solve the equation is the Kaczmarz's

Method by iteration and the method is applied to many tomography problems. Meanwhile, we shall give a direct and explicit representation of the inverse of (4.2). So, we are interested in applications of our direct solutions for (4.2) to the related many problems.

By the Gauss fundamental procedure, we can obtain

Lemma 4.1. *Let A be a matrix of type $m \times n$ and with rank p. Then, there exist vectors $\{\boldsymbol{b}_h\}_{h=1}^p$ in $\mathbb{R}^m$ such that the vectors $\{\boldsymbol{b}_h^\top A\}_{h=1}^p$ are linearly independent row vectors.*

Lemma 4.2. *Let A and B be matrices of types $m \times n$ and $l \times n$, respectively. Assume that A has rank p and the augmented matrix*

$$\begin{pmatrix} A \\ B \end{pmatrix} \tag{4.4}$$

has rank q. Then, there are vectors $\{\boldsymbol{b}_h\}_{h=1}^p$ in $\mathbb{R}^m$ and $\{\boldsymbol{d}_h\}_{h=1}^{q-p}$ in $\mathbb{R}^l$ such that the row vectors $\boldsymbol{b}_1^\top A$, $\boldsymbol{b}_2^\top A$, $\ldots$, $\boldsymbol{b}_p^\top A$, $\boldsymbol{d}_1^\top B$, $\boldsymbol{d}_2^\top B$, $\ldots$, $\boldsymbol{d}_{q-p}^\top B$ are linearly independent.

Lemma 4.3. *Let $A_1^{j_1}$, $A_2^{j_2}, \ldots$, $A_r^{j_r}$ be given matrices of types $j_1 \times n$, $j_2 \times n, \ldots$, $j_r \times n$, respectively. Let the rank of the matrix*

$$\begin{pmatrix} A_1^{j_1} \\ A_2^{j_2} \\ \vdots \\ A_s^{j_s} \end{pmatrix} \tag{4.5}$$

be p_s, for $s = 1, 2, \ldots, r$. Let $p_r = n$ but $p_{r-1} < n$. Then there are p_1 vectors $\boldsymbol{b}_1^{j_1}$, $\boldsymbol{b}_2^{j_1}, \ldots$, $\boldsymbol{b}_{p_1}^{j_1} \in \mathbb{R}^{j_1}$, $p_2 - p_1$ vectors $\boldsymbol{b}_{p_1+1}^{j_2}$, $\boldsymbol{b}_{p_1+2}^{j_2}, \ldots$, $\boldsymbol{b}_{p_2}^{j_2} \in \mathbb{R}^{j_2}$, $\ldots$, $(n - p_{r-1})$ vectors $\boldsymbol{b}_{p_{r-1}+1}^{j_r}$, $\boldsymbol{b}_{p_{r-1}+2}^{j_r}, \ldots$, $\boldsymbol{b}_n^{j_r} \in \mathbb{R}^{j_r}$ such that the vectors $(\boldsymbol{b}_{s_1}^{j_1})^\top A_1^{j_1}$, $(\boldsymbol{b}_{s_2}^{j_2})^\top A_2^{j_2}$, $\ldots$, $(\boldsymbol{b}_{s_r}^{j_r})^\top A_r^{j_r}$, where $(s_1 = 1, 2, \ldots, p_1)$, $(s_2 = p_1 + 1, p_1 + 2, \ldots, p_2)$, $\ldots$, $(s_r = p_{r-1} + 1, p_{r-1} + 2, \ldots, n)$ are linearly independent.

Then, we obtain the direct inversion formula for (4.2)

Theorem 4.1. *Let $A_1^{j_1}$, $A_2^{j_2}, \ldots$, $A_r^{j_r}$ be given matrices with the properties as in Lemma 4.3 and let $\boldsymbol{b}_{s_1}^{j_1} \in \mathbb{R}^{j_1}$, $\boldsymbol{b}_{s_2}^{j_2} \in \mathbb{R}^{j_2}$, $\ldots$, $\boldsymbol{b}_{s_r}^{j_r} \in \mathbb{R}^{j_r}$, for $(s_1 = 1, 2, \ldots, p_1)$, $(s_2 = p_1 + 1, p_1 + 2, \ldots, p_2)$, $\ldots$, $(s_r = p_{r-1} + 1, p_{r-1} + 2, \ldots, n)$ satisfying Lemma 4.3. Let for $\boldsymbol{x} \in \mathbb{R}^n$, $A_k^{j_k} \boldsymbol{x} = \boldsymbol{y}_k^{j_k}$, for $k = 1, 2, \ldots, r$, be* *given. Then*

$$\boldsymbol{x} = \begin{pmatrix} (\boldsymbol{b}_{s_1}^{j_1})^\top A_1^{j_1} \\ \vdots \\ (\boldsymbol{b}_{s_r}^{j_r})^\top A_r^{j_r} \end{pmatrix}^{-1} \begin{pmatrix} (\boldsymbol{b}_{s_1}^{j_1})^\top \boldsymbol{y}_1^{j_1} \\ \vdots \\ (\boldsymbol{b}_{s_r}^{j_r})^\top \boldsymbol{y}_r^{j_r} \end{pmatrix},$$

$$1 \le s_1 \le p_1, \quad p_1 + 1 \le s_2 \le p_2, \quad \ldots, \quad p_{r-1} + 1 \le s_r \le n. \tag{4.6}$$

We can also obtain one more constructive and direct solution of (4.2).

Theorem 4.2. *Let $A_1^{j_1}, A_2^{j_2}, \ldots, A_r^{j_r}$ be given matrices of types $j_1 \times n$, $j_2 \times n, \ldots, j_r \times n$, respectively such that the rank of the augmented matrix*

$$\begin{pmatrix} A_1^{j_1} \\ A_2^{j_2} \\ \vdots \\ A_r^{j_r} \end{pmatrix} \tag{4.7}$$

is n. For $\boldsymbol{x} \in \mathbb{R}^n$, let $A_k^{j_k}\boldsymbol{x} = \boldsymbol{y}_k^{j_k}$, for $k = 1, 2, \ldots, r$, be given. Then there are n vectors $\boldsymbol{b}_{1k}^{j_k}$, $\boldsymbol{b}_{2k}^{j_k}$, $\ldots$, $\boldsymbol{b}_{nk}^{j_k} \in \mathbb{R}^{j_k}$ for $k = 1, 2, \ldots, r$, such that

$$\boldsymbol{x} = \begin{pmatrix} \sum_{k=1}^{r} (\boldsymbol{b}_{1k}^{j_k})^\top A_k^{j_k} \\ \sum_{k=1}^{r} (\boldsymbol{b}_{2k}^{j_k})^\top A_k^{j_k} \\ \vdots \\ \sum_{k=1}^{r} (\boldsymbol{b}_{nk}^{j_k})^\top A_k^{j_k} \end{pmatrix}^{-1} \begin{pmatrix} \sum_{k=1}^{r} (\boldsymbol{b}_{1k}^{j_k})^\top \boldsymbol{y}_k^{j_k} \\ \sum_{k=1}^{r} (\boldsymbol{b}_{2k}^{j_k})^\top \boldsymbol{y}_k^{j_k} \\ \vdots \\ \sum_{k=1}^{r} (\boldsymbol{b}_{nk}^{j_k})^\top \boldsymbol{y}_k^{j_k} \end{pmatrix}. \tag{4.8}$$

In Theorems 4.1 and 4.2, we assumed that the rank of the augmented matrix is n, a full rank for $\boldsymbol{x} \in \mathbb{R}^n$, however, in general, we can obtain a generalized inverse with minimum norm, as in (II) and (III).

5. GENERALIZED PYTHAGOREAN THEOREMS

Let $\{\boldsymbol{e}_i\}_{i=1}^n$ be a basis of $\mathbb{R}^n$. For $i = 1, 2, \ldots, n$, we define the maps $L_i : \mathbb{R}^n \to \mathbb{R}^n$ by $L_i(\boldsymbol{x}) = \langle \boldsymbol{x}, \boldsymbol{e}_i \rangle \boldsymbol{e}_i$, and the maps $M_i : \mathbb{R}^n \to \mathbb{R}^n$ by $M_i(\boldsymbol{x}) = \boldsymbol{x} - \langle \boldsymbol{x}, \boldsymbol{e}_i \rangle \boldsymbol{e}_i$, then clearly L_i and M_i are linear maps and for $\boldsymbol{b} \in \mathbb{R}^n$, $\langle L_i(\boldsymbol{x}), \boldsymbol{b} \rangle = \langle \boldsymbol{x}, L_i(\boldsymbol{b}) \rangle$ and $\langle M_i(\boldsymbol{x}), \boldsymbol{b} \rangle = \langle \boldsymbol{x}, M_i(\boldsymbol{b}) \rangle$. Hence, they are self adjoint. Now, we define the matrices

$$P_x = [L_1(\boldsymbol{x}) \quad L_2(\boldsymbol{x}) \quad \cdots \quad L_n(\boldsymbol{x})] \tag{5.1}$$

and

$$Q_x = [M_1(\boldsymbol{x}) \quad M_2(\boldsymbol{x}) \quad \cdots \quad M_n(\boldsymbol{x})]. \tag{5.2}$$

Then the maps $L : \mathbb{R}^n \to M_n(\mathbb{R})$ and $M : \mathbb{R}^n \to M_n(\mathbb{R})$ defined respectively by $L(\boldsymbol{x}) = P_x$ and $M(\boldsymbol{x}) = Q_x$ are injective linear maps. The problems are (1) to find $\boldsymbol{x}$ and its norm from the data $L_i(\boldsymbol{x})$ $(i = 1, 2, \ldots, n)$ and (2) to find $\boldsymbol{x}$ and its norm from the data $M_i(\boldsymbol{x})$ $(i = 1, 2, \ldots, n)$. Consider the problems in their geometrical meanings and from the viewpoint of the Pythagorean theorem.

Then we can see, directly

Lemma 5.1. *For $\boldsymbol{x}, \boldsymbol{b} \in \mathbb{R}^n$, (i) $P_x^\top \boldsymbol{b} = P_b^\top \boldsymbol{x}$ and (ii) $Q_x^\top \boldsymbol{b} = Q_b^\top \boldsymbol{x}$.*

Lemma 5.2. *Let $\boldsymbol{b} \in \mathbb{R}^n$. Then, P_b is nonsingular if and only if $\langle \boldsymbol{b}, \boldsymbol{e}_j \rangle \neq 0$ for $j = 1, 2, \ldots, n$. For $\boldsymbol{b} = \sum_{i=1}^n \alpha_i \boldsymbol{e}_i$ and $a_i = \langle \boldsymbol{b}, \boldsymbol{e}_i \rangle$, Q_b is nonsingular if and only if $a_j \neq 0$ for all j and $\sum_{i=1}^n \alpha_i / a_i \neq 1$.*

Here, we note that if $\{\boldsymbol{e}_i\}_{i=1}^n$ is an orthonormal basis, then $\sum_{i=1}^n \alpha_i / a_i = n$, i.e., the assumption $\sum_{i=1}^n \alpha_i / a_i \neq 1$ automatically holds but in general, $\{\boldsymbol{b} - \langle \boldsymbol{b}, \boldsymbol{e}_j \rangle \boldsymbol{e}_j\}_{j=1}^n$ may be linearly dependent, for $n > 2$ under the condition $\langle \boldsymbol{b}, \boldsymbol{e}_j \rangle \neq 0$ for $j = 1, 2, \ldots, n$. So, in Lemma 5.2, we see that the assumption $\sum_{i=1}^n \alpha_i / a_i \neq 1$ is, in general, needed for the regularity of Q_b.

Then, we obtain the generalizations of the Pythagorean theorem

Theorem 5.1. *Let $\boldsymbol{b} \in \mathbb{R}^n$ as in Lemma 5.2. Then, the inverse linear map $L^{-1} : L(\mathbb{R}^n) \to \mathbb{R}^n$ is obtained by the formula*

$$\boldsymbol{x} = (PP_b^{-1})^\top \boldsymbol{b}, \quad \textit{for} \quad P \in L(\mathbb{R}^n) \tag{5.3}$$

and

$$\|\boldsymbol{x}\|^2 = (\boldsymbol{b}^\top P) K_1^{-1} (P^\top \boldsymbol{b}), \tag{5.4}$$

where ij-th entry of K_1 is the product $\langle \boldsymbol{b}, \boldsymbol{e}_i \rangle \langle \boldsymbol{b}, \boldsymbol{e}_j \rangle \langle \boldsymbol{e}_i, \boldsymbol{e}_j \rangle$.

Theorem 5.2. *Let $\boldsymbol{b} \in \mathbb{R}^n$ as in Lemma 5.2. Then, the inverse linear map $M^{-1} : M(\mathbb{R}^n) \to \mathbb{R}^n$ is obtained by the formula*

$$\boldsymbol{x} = (QQ_b^{-1})^\top \boldsymbol{b} \qquad \textit{for} \quad Q \in M(\mathbb{R}^n) \tag{5.5}$$

and

$$\|\boldsymbol{x}\|^2 = (\boldsymbol{b}^\top Q) K_2^{-1} (Q^\top \boldsymbol{b}), \tag{5.6}$$

where ij-th entry of K_2 is the inner product $\langle \boldsymbol{b} - \langle \boldsymbol{b}, \boldsymbol{e}_i \rangle \boldsymbol{e}_i, \boldsymbol{b} - \langle \mathbf{b}, \boldsymbol{e}_j \rangle \boldsymbol{e}_j \rangle$.

The above theorems can be generalized as follows:

Theorem 5.3. *Assume that $L_i : \mathbb{R}^n \to \mathbb{R}^n$, $(i = 1, 2, \ldots, p)$ are linear operators and $M_j : \mathbb{R}^n \to \mathbb{R}$,$(j = 1, 2, \ldots, q)$ are linear functionals such that $p + q = n$. Then by the Riesz representation theorem, there exist unique vectors $\boldsymbol{c}_j$ such that $M_j(\boldsymbol{x}) = \langle \boldsymbol{x}, \boldsymbol{c}_j \rangle$ $(j = 1, 2, \ldots, q)$. If there is a vector $\boldsymbol{b}$ such that the set $\{L_1^*(\boldsymbol{b}), L_2^*(\boldsymbol{b}), \ldots, L_p^*(\boldsymbol{b}), \boldsymbol{c}_1, \boldsymbol{c}_2, \ldots, \boldsymbol{c}_q\}$ is linearly independent. Then,*

$$\boldsymbol{x} = \left([L_1^*(\boldsymbol{b})\ L_2^*(\boldsymbol{b})\ \ldots\ L_p^*(\boldsymbol{b})\ \boldsymbol{c}_1\ \boldsymbol{c}_2\ \ldots\ \boldsymbol{c}_q]^\top \right)^{-1} \begin{pmatrix} \langle \boldsymbol{b}, L_1(\boldsymbol{x}) \rangle \\ \langle \boldsymbol{b}, L_2(\boldsymbol{x}) \rangle \\ \vdots \\ \langle \boldsymbol{b}, L_p(\boldsymbol{x}) \rangle \\ M_1(\boldsymbol{x}) \\ M_2(\boldsymbol{x}) \\ \vdots \\ M_q(\boldsymbol{x}) \end{pmatrix}. \tag{5.7}$$

6. GENERALIZED SOLUTIONS

In order to represent our generalized solutions explicitly, we shall consider bounded linear operators on a reproducing kernel Hilbert space. So, we consider the Hilbert space H_K on E stated in Section 1. We consider H as H_K in Section 3. We assume that the direct integral

$$\boldsymbol{H} = \int_{\boldsymbol{\Lambda}}^{\oplus} \boldsymbol{H}_\lambda \, d\mu(\lambda) \tag{6.1}$$

of the Hilbert spaces $\boldsymbol{H}_\lambda$ on $\boldsymbol{\Lambda}$ converges with a σ finite positive measure $d\mu$ on $\boldsymbol{\Lambda}$. We assume that the bounded linear operators L_λ in (3.2) are bounded on H_K into $\boldsymbol{H}$ in the sense:

$$\int_{\boldsymbol{\Lambda}} \|L_\lambda f\|_{\boldsymbol{H}_\lambda}^2 \, d\mu(\lambda) \leq M \|f\|_{H_K}^2 \tag{6.2}$$

for some constant $M \geq 0$.

In this setting, we consider the extremal problem:

$$\inf_{f \in H_K} \int_{\boldsymbol{\Lambda}} \|L_\lambda f - g(\lambda)\|_{\boldsymbol{H}_\lambda}^2 \, d\mu(\lambda), \tag{6.3}$$

which gives a generalized solution for the equations

$$L_\lambda f = g(\lambda) \quad \text{on} \quad H_K \quad \text{and} \quad \boldsymbol{H}_\lambda. \tag{6.4}$$

We shall write the operators $\{L_\lambda\}$ as L from H_K into $\mathbf{H}$ in the sense (6.2). Let L^* be the adjoint operator of L from $\boldsymbol{H}$ into H_K. We form the positive matrix

$$k(p,q) = (L^*LK(\cdot,q), L^*LK(\cdot,p))_{H_K} \quad \text{on} \quad E \times E. \tag{6.5}$$

Then, we obtain

Theorem 6.1. *For a function $g \in \boldsymbol{H}$, there exists a function $\tilde{f}$ in H_K such that*

$$\inf_{f \in H_K} \int_{\boldsymbol{\Lambda}} \|L_\lambda f - g(\lambda)\|^2_{\boldsymbol{H}_\lambda} \, d\mu(\lambda) = \int_{\boldsymbol{\Lambda}} \|L_\lambda \tilde{f} - g(\lambda)\|^2_{\boldsymbol{H}_\lambda} \, d\mu(\lambda) \tag{6.6}$$

if and only if, for the RKHS H_k

$$L^* g \in H_k. \tag{6.7}$$

Furthermore, if there exist the best approximations $\tilde{f}$ satisfying (6.6), then there exists a unique extremal function $\check{f}$ with the minimum norm in H_K, and this function is expressible in the form

$$\check{f}(p) = (L^*g, L^*LK(\cdot,p))_{H_k} \quad \text{on} \quad E. \tag{6.8}$$

In this theorem, note that

$$(L^*g)(p) = (L^*g, K(\cdot,p))_{H_K} = (g, LK(\cdot,p))_{\boldsymbol{H}}; \tag{6.9}$$

that is, the adjoint operator L^* is expressible in terms of $g, L, K(\cdot,p)$ and $\boldsymbol{H}$.

For some proof of this theorem, we can apply the argument in [3].

As a simple example, we shall consider the space H_K on $[0,\infty)$ for $K(x,y) = \min\{x,y\}$. This space is composed of all absolutely continuous real-valued functions on $[0,\infty)$ satisfying $f(0) = 0$ and equipped with the norm

$$\|f\|^2_{H_K} = \int_0^\infty f'(x)^2 \, dx. \tag{6.10}$$

As a space $\boldsymbol{H}$ we consider the space $L_2((0,\infty), e^{-\lambda} d\lambda)$ and a bounded linear operator L:

$$L_\lambda f = \int_0^\lambda f(\xi) \, d\xi \tag{6.11}$$

from H_K into $\boldsymbol{H}$. Then, the adjoint operator L^* from $\boldsymbol{H}$ into H_K is given by

$$(L^*g)(x) = \frac{1}{2}\int_0^x g(\lambda)\lambda^2 e^{-\lambda}\,d\lambda + \int_x^\infty g(\lambda)\left[x\lambda - \frac{1}{2}x^2\right]e^{-\lambda}\,d\lambda \tag{6.12}$$

and we can discuss the problem, for any $g \in \boldsymbol{H}$,

$$\inf_{f\in H_K} \int_0^\infty \left|\int_0^\lambda f(\xi)\,d\xi - g(\lambda)\right|^2 e^{-\lambda}\,d\lambda. \tag{6.13}$$

We can give a complete solution for the problem. We would like to discuss those concrete problems in separate papers.

We are interested in some concrete results for typical problems such as generalized solutions for ordinary differential equations in connection with reproducing kernels, Green's functions and the related completeness in (6.8).

REFERENCES

1. N. Aronszajn, Theory of reproducing kernels. *Trans. Amer. Math. Soc.* (1950) **68**, 337–404.

2. D.-W. Byun and S. Saitoh, Approximation by the solutions of the heat equation. *J. Approximation Theory* (1994) **78**, 226–238.

3. D.-W. Byun and S. Saitoh, Best approximation in reproducing kernel Hilbert spaces. In: *Proc. of the 2th International Colloquium on Numerical Analysis.* VSP-Holland, 1994, 55–61.

4. N. Hayashi, Analytic function spaces and their applications to nonlinear evolution equations. In: *Analytic Extension Formulas and their Applications.* Kluwer Academic Publishers, 2001, 59–86.

5. J. R. Higgins, A sampling principle associated with Saitoh's fundamental theory of linear transformations. *Analytic Extension Formulas and their Applications.* Kluwer Academic Publishers, 2001, 73–86.

6. F. Natterer, *The mathematics of Computerized Tomography.* SIAM in Applied Mathematics, **32**. SIAM, Philadelphia, 2001.

7. Th. M. Rassias and S. Saitoh, The Pythagorean theorem and linear mappings. *PanAmerican Math. J.* (2002) **12**, 1–10.

8. S. Saitoh, Hilbert spaces induced by Hilbert space valued functions. *Proc. Amer. Math. Soc.* (1983) **89**, 74–78.

9. S. Saitoh, Theory of reproducing kernels and its applications. *Pitman Research Notes in Mathematics Series* (1988) **189**.

10. S. Saitoh, Interpolation problems of Pick—Nevanlinna type. *Pitman Research Notes in Mathematics Series* (1989) **212**, 253–262.

11. S. Saitoh, Representations of the norms in Bergman—Selberg spaces on strips and half planes. *Complex Variables* (1992) **19**, 231–241.

12. S. Saitoh, One approach to some general integral transforms and its applications. *Integral Transforms and Special Functions* (1995) **3**, 49–84.

13. S. Saitoh, Natural norm inequalities in nonlinear transforms. *General Inequalities* (1997) **7**, 39–52.

14. S. Saitoh, Representations of inverse functions. *Proc. Amer. Math. Soc.* (1997) **125**, 3633–3639.

15. S. Saitoh, Integral transforms, reproducing kernels and their applications. *Pitman Research Notes in Mathematics Series* (1997) **369**.

16. S. Saitoh, Nonlinear transforms and analyticity of functions. In: *Nonlinear Mathematical Analysis and Applications.* Hadronic Press, Palm Harbor, 1998, 223–234.

17. S. Saitoh, Various operators in Hilbert space induced by transforms. *Intern. J. Appl. Math.* (1999) **1**, 111–126.

18. S. Saitoh, Applications of the general theory of reproducing kernels. In: *Reproducing Kernels and their Applications.* Kluwer Academic Publishers, 1999, 165–188.

19. S. Saitoh and M. Yamamoto, Integral transforms involving smooth functions. In: *Reproducing Kernels and their Applications.* Kluwer Academic Publishers, 1999, 149–164.

20. S. Saitoh, Linear integro-differential equations and the theory of reproducing kernels. In: *Volterra Equations and Applications.* C. Corduneanu and I. W. Sandberg (Eds). Gordon and Breach Science Publishers, Amsterdam, 2000.

21. S. Saitoh, Analytic extension formulas, integral transforms and reproducing kernels. In: *Analytic Extension Formulas and their Applications.* Kluwer Academic Publishers, , 2001, 207–232.

22. S. Saitoh, Applications of the reproducing kernel theory to inverse problems. *Comm. Korean Math. Soc.* (2001) **16**, 371–383.

23. S. Saitoh and M. Mori, Representations of analytic functions in terms of local values by means of the Riemann mapping function. *Complex Variables* (2001) **45**, 387–393.

24. S. Saitoh, Principle of telethoscope. *Proceedings of the Intern. Workshop Functional-Analytic and Complex Methods, their Interaction and Applications to Partial Differential Equations.* Graz, Austria, 12–16 February, 2001. World Scientific, 2001, 101–117.

25. S. Saitoh, Weighted L_p-norm inequalities in convolutions. In: *Survey on Classical Inequalities.* Kluwer Academic Publishers, 2000, 225–234.

26. S. Saitoh, V. K. Tuan, and M. Yamamoto, Reverse L_p-norm inequalities in convolutions and stability in inverse problems. *J. of Inequalities in Pure and Applied Mathematics* (2000) **1**, No. 7.

27. S. Saitoh, V. K. Tuan, and M. Yamamoto, Reverse convolution inequalities and applications to inverse heat source problems. (In preparation).

28. M. Asaduzzaman and S. Saitoh, Inverses of a family of matrices and generalizations of Pythagorean theorem. (In preparation).

Ill-Posed and Non-Classical Problems of
Mathematical Physics and Analysis, Samarkand, 2000, pp. 143–172
M. M. Lavrent'ev and S. I. Kabanikhin (Eds)

Cauchy problem for the Helmholtz equation

Sh. YARMUKHAMEDOV and I. YARMUKHAMEDOV

1. NOTATIONS AND SETTINGS OF THE PROBLEM

Let E^3 be a real Euclidean space, $y = (y_1, y_2, y_3)$, $x = (x_1, x_2, x_3) \in E^3$, $\alpha^2 = (y_1 - x_1)^2 + (y_2 - x_2)^2$, $r^2 = |y - x|^2 = \alpha^2 + (y_3 - x_3)^2$, $s = \alpha^2$, D be a bounded one-connected domain in E^3 with the boundary $\partial\Delta$ consisted of a compact connected part T of the domain $y_3 = 0$ and the smooth Lyapunov surface S lying in the half-space $y_3 > 0$, $\partial D = S \bigcup T$, $\overline{D} = D \bigcup \partial D$; $d/(dn)$ be the differentiation operator along the external normal to S (or to T); $\lambda = \lambda(y)$ be a complex-valued function set in D. Let

$$\lambda(y) = \lambda_1(y) + i\lambda_2(y),$$

where $\lambda_1(y) = \lambda_1 = \operatorname{Re}\lambda(y)$, $\lambda_2(y) = \lambda_2 = \operatorname{Im}\lambda(y)$; $C^k(D)$ be the space consisted of k times continuously differentiable complex-valued functions defined in D and Δ_y be the three-dimensional Laplace operator

$$\Delta_y = \frac{\partial^2}{\partial y_1^2} + \frac{\partial^2}{\partial y_2^2} + \frac{\partial^2}{\partial y_3^2}.$$

The problem (Cauchy). Suppose $U(y) \in C^2(D)$ and

$$\Delta U(y) - \lambda^2(y)U(y) = 0, \qquad y \in D, \tag{1.1}$$

$$U(y) = f(y), \qquad \frac{\partial U}{\partial n}(y) = g(y), \quad y \in S, \tag{1.2}$$

where $f(y)$ and $g(y)$ are given functions in S from the classes $C^1(S)$ and $C(S)$ respectively. We need to recover $U(y)$ in D.

Problem (1.1)–(1.2) is not solvable for arbitrary $\lambda(y)$, $f(y)$, and $g(y)$. If S is an analytical surface, the function $\lambda(y)$ is analytical in $D \cup S$ and the functions $f(y)$ and $g(y)$ are analytical in S and may be extended analytically into D, then the solution exists, is unique but is unstable. Therefore, problem (1.1)–(1.2) is an ill-posed problem.

We suppose that $\lambda(y) \in C^2(\overline{D})$ and the solution of the problem exists (then it is unique) and belongs to the class $C^2(D) \cap C^1(\overline{D})$. Under this assumption we shall construct the solution if the Cauchy date $f(y)$ and $g(y)$ are given exactly. If, moreover, the solution satisfies the condition in the plane part of the boundary

$$|U(y)| + \left|\frac{d}{dn} U(y)\right| \leq M, \qquad y \in T, \tag{1.3}$$

where M is a given positive number, then we shall construct the regularized solution, when, instead of $f(y)$ and $g(y)$, their continuous approximations $f_\delta(y)$ and $g_\delta(y)$ are given in S:

$$\sup_S |f(y) - f_\delta(y)| < \delta, \qquad \sup_S |g(y) - g_\delta(y)| < \delta. \tag{1.4}$$

The formulas written below are based on the setting and the method of Lavrent'ev, M. M. [1,2]. The essence of this method is the explicit construction of the Levi function. This function depends on the positive parameter δ and vanishes with its derivatives in the plane part T of the boundary where $y_3 = 0$ when the parameter tends to infinity in the case when the pole x of the Levi function lies in the domain D. The Levi function with the above properties is called the Carleman function for problem (1.1)–(1.2) [3] (the definition of the Carleman function we had given suited for our problem).

We shall construct the Carleman function explicitly. If $\lambda(y) = \text{const}$, then the Carleman function will be the fundamental solution of equation (1.1) and the solution of problem (1.1), (1.2) and the regularization will be written explicitly in the form of the difference of the potentials of simple and double layers. If the function $\lambda(y)$ is not a constant and belongs to the class $C^2(\overline{D})$, then the solution of the integral Fredholm equation of the second kind with the kernel dependent on a positive parameter. The series consisted of the iterated kernels converges absolutely and uniformly for the parameter δ sufficiently large, and the solution is expressed via the resolvents of the integral equation. The formula obtained below allows us to construct easily the solution of problem (1.1)–(1.2) in both cases: when the Cauchy data are given explicitly and the regularized solution, when the Cauchy data are given approximately.

When the number of variables is equal to two, the Cauchy problem for the Laplace equation was investigated first by Carleman in 1926 for the bounded domain with the boundary consisted from the parts of the rays with the common origin and the smooth curve lying inside the angle [4]. Carleman had founded the formula of recovery the holomorphic function in the points of the domain lying in the bisectrix of the angle given the values of this function in the boundary curve. The generalization of the Carleman formula (both one-dimensional and multidimensional) can be found in the monograph [5]. There is also the detailed bibliography on this problem.

To obtain the extension formula Carleman had adapted the Cauchy formula to this problem by introduction a new idea. He had replaced in the Cauchy integral the kernel by the generalized kernel dependent on a positive parameter which is holomorphic inside the angle except for a simple pole. This kernel vanishes in the sides of the angle when the parameter tends to infinity. The recovery formula being found by Carleman is unstable. The further investigations of M. M. Lavrent'ev had shown that the introduction of new positive parameter here was stipulated by the essence of the problem and for special choice of the parameter the stable formulas of extension in the class of bounded holomorphic functions may be constructed.

The problem of construction of multidimensional analog of the Carleman formula on the basis of the classical Green formula had led to the notion of the Carleman function [1]. The explicit form of the Carleman function and the multidimensional analog of the Carleman formula connected with it was found in [6–10]. In the case when the number of variables is equal to two and λ is a positive constant, the Carleman function and the regularization of the Cauchy problem for the Helmholtz equation was constructed explicitly in [10]. For many variables and for complex λ this function was constructed in [8].

2. CONSTRUCTION OF THE CARLEMAN FUNCTION AND THE INTEGRAL EQUATION OF THE SECOND KIND

Suppose that $\lambda(y) \in C^2(\overline{D})$ and σ is a positive number. Denote

$$K(w) = e^{\sigma w^2}, \qquad w = i\sqrt{u^2+\alpha^2} + y_3.$$

The function $\Phi_\sigma(x, y)$ for $\alpha > 0$ we define by the equality

$$-2\pi^2 K(x_3)\Phi_\sigma(x,\, y) = \int_0^\infty \operatorname{Im}\left[\frac{K(w)}{w - x_3}\right]\frac{\cos\lambda u\, du}{\sqrt{u^2+\alpha^2}}. \tag{2.1}$$

If we we separate the imaginary part, then $\Phi_\sigma(x,y)$ will be as follows:

$$2\pi^2\Phi_\sigma(x,\, y) = \int_0^\infty F_\sigma(x,\, y,\, u)\cos\lambda u\, du, \tag{2.2}$$

where

$$F_\sigma(x,y,u) = \frac{\varphi_\sigma(x,\, y\, u)}{u^2+r^2}\, e^{\sigma(y_3^2-x_3^2)-\sigma\alpha^2-\sigma u^2},$$

$$\varphi_\sigma(x,y,u) = \cos\tau\sqrt{u^2+\alpha^2} - (y_3-x_3)\frac{\sin\tau\sqrt{u^2+\alpha^2}}{\sqrt{u^2+\alpha^2}},$$

$$\tau = 2\sigma y_3. \tag{2.3}$$

Convergence of the improper integral in the right-hand side of equalities (2.1) and (2.2) is guaranteed by the multiplier $e^{-\sigma u^2}$.

The basic lemma. *The function defined by equality* (2.1) (*or* (2.2)) *is represented in the form*

$$\Phi_\sigma(x,y) = \frac{e^{-\lambda(y)r}}{4\pi r} + G_\sigma(x,\, y) \tag{2.4}$$

where $G_\sigma(x,y)$ is the function from the class $C^2(\overline{D})$with respect to the variable y including the point $y=x$.

Proof. Divide the interval of integration in (2.1) into two parts: $(0,\infty) = (0,1)\cup(1,\infty)$. The second integral we denote as $g_1(y,x)$. Since $1\le u<\infty$, the function $g_1(y,x)$ belongs to the class $C^2(\overline{D})$ with respect to the variable y including the point $y=x$. The first integral we denote by $g_2(x,y)$. Transform this integral as follows:

$$g_2(y,x) = \frac{1}{K(x_3)}\int_0^1 \operatorname{Im}\left[\frac{K(w)-K(x_3)}{w-x_3}\right]\frac{\cos\lambda u}{\sqrt{u^2+\alpha^2}}\,du + \int_0^1 \operatorname{Im}\left[\frac{1}{w-x_3}\right]\frac{\cos\lambda u\, du}{\sqrt{u^2+\alpha^2}},$$

where

$$\operatorname{Im}\frac{1}{w-x_3} = -\frac{\sqrt{u^2+\alpha^2}}{u^2+r^2}.$$

Since the entire function

$$K_1(w) = \frac{K(w)-K(x_3)}{w-x_3}$$

is real for real w, then, from the expansion

$$\frac{1}{\sqrt{u^2+\alpha^2}}\,\mathrm{Im}[K_1(w)] = \sum_{n=0}^{\infty}(-1)^n \frac{\mathrm{Im}\,K_1^{(2n+1)}(x_3)}{(2n+1)!}(u^2+\alpha^2)^n$$

it follows that the first term is the function from the class $C^2(\overline{D})$. The second integral from $g_2(y,x)$ we denote by $\psi(x,y)$. Taking into account the equality

$$\cos u = \sum_0^{\infty}(-1)^n \frac{u^{2n}}{(2n)!},$$

we transform it as follows:

$$\begin{aligned}\Psi(x,y) &= -\int_0^1 \frac{\cos\lambda u}{u^2+r^2}\,du = -\sum_{n=0}^{\infty}\frac{(-1)^n\lambda^{2n}}{(2n)!}\int_0^1 \frac{u^{2n}}{u^2+r^2}\,du \\ &= -\int_0^1 \frac{du}{u^2+r^2} + \frac{\lambda^2}{2}\int_0^1 \frac{u^2du}{u^2+r^2} - \sum_{n=2}^{\infty}\frac{(-1)^n\lambda^{2n}}{(2n)!}\int_0^1 \frac{u^{2n}\,du}{u^2+r^2}.\end{aligned}$$

The first integral is equal to

$$\frac{\pi}{2r} - \int_1^{\infty}\frac{du}{u^2+r^2},$$

and the second integral is equal to

$$\frac{\lambda^2}{2} - \frac{r^2\lambda^2}{2}\Big(\frac{\pi}{2r} - \int_1^{\infty}\frac{du}{u^2+r^2}\Big).$$

Therefore, the function $\psi(x,y)$ has the form

$$\begin{aligned}\Psi(x,y) &= -\frac{\pi}{2r} + \frac{\lambda^2}{2} - \frac{\pi r\lambda^2}{4} \\ &\quad + \Big(1+\frac{r^2\lambda^2}{2}\Big)\int_1^{\infty}\frac{du}{u^2+r^2} - \sum_{n=2}^{\infty}\frac{(-1)^n\lambda^{2n}}{(2n)!}\int_0^1\frac{u^{2n}\,du}{u^2+r^2}.\end{aligned}$$

Since

$$-\frac{\pi}{2r} - \frac{\pi r\lambda^2}{4} = -\frac{\pi}{2}\frac{e^{-\lambda r}}{r} - \frac{\pi\lambda}{2} + \frac{\pi}{2}\sum_{n=3}^{\infty}(-1)^n\frac{\lambda^n r^{n-1}}{n!},$$

we obtain

$$g_2(y,x) = -\pi e^{-\lambda r}/(2r) + g_3(y,x),$$

where $g_3(y,x)$ is a function from the class $C^2(\overline{D})$with respect to the variable y including the point $y = x$. Setting

$$G_\sigma(x,y) = -(g_1(x,y) + g_2(x,y))/(2\pi^2),$$

we obtain the lemma statement. □

Lemma 2.1. *For $y \neq x$ the following equality holds*

$$\frac{1}{2\pi^2}\int_0^\infty \Delta_y\Big[F_\sigma(x,y,u)\Big]\cos\lambda u\,du = \lambda^2\Phi_\sigma(x,y). \tag{2.5}$$

Proof. Denote

$$\Psi_\sigma(w) = -\frac{1}{2\pi^2 K(x_3)}\frac{K(w)}{w-x_3}.$$

Then

$$-\frac{1}{2\pi^2}\frac{1}{K(x_3)}\operatorname{Im}\frac{K(w)}{w-x_3} = \frac{\Psi_\sigma(w)-\Psi_\sigma(\overline{w})}{2i}, \quad \overline{w} = -i\sqrt{u^2+\alpha^2}+y_3$$

and

$$\frac{1}{2\pi^2}\Delta_y\Big[F_\sigma(x,y,u)\Big] = \frac{1}{2i}\Delta_y\Big[\frac{\Psi_\sigma(w)-\Psi_\sigma(\overline{w})}{\sqrt{u^2+\alpha^2}}\Big].$$

The function under the Laplace operator Δ_y depends on $s = \alpha^2$ and y_3, i.e. is the function dependent on the point $t = (s, y_3)$. The simple computation shows that the Laplace operator in the coordinates of the point t has the form

$$\Delta_t = 4s\frac{d^2}{ds^2} + 4\frac{d}{ds} + \frac{d^2}{dy_3^2}; \tag{2.6}$$

therefore,

$$\begin{aligned}\frac{1}{2\pi^2}&\int_0^\infty \Delta_y[F_\sigma(x,y,u)]\cos\lambda u\,du\\ &= \frac{1}{2i}\Big[\int_0^\infty \cos\lambda u\Delta_t\Big(\frac{\Psi_\sigma(w)}{\sqrt{u^2+\alpha^2}}\Big)du - \int_0^\infty \cos\lambda u\Delta_t\Big(\frac{\Psi_\sigma(\overline{w})}{\sqrt{u^2+\alpha^2}}\Big)du\Big].\end{aligned}$$

Now, the formula (2.5) follows from the equalities

$$\begin{aligned}\int_0^\infty \cos\lambda u\Delta_t\Big(\frac{\Psi_\sigma(w)}{\sqrt{u^2+\alpha^2}}\Big)du &= \lambda^2\int_0^\infty \frac{\Psi_\sigma(w)}{\sqrt{u^2+\alpha^2}}\cos\lambda u\,du,\\ \int_0^\infty \cos\lambda u\Delta_t\Big(\frac{\Psi_\sigma(\overline{w})}{\sqrt{u^2+\alpha^2}}\Big)du &= \lambda^2\int_0^\infty \frac{\Psi_\sigma(\overline{w})}{\sqrt{u^2+\alpha^2}}\cos\lambda u\,du\end{aligned} \tag{2.7}$$

and formula (2.1). We prove now the first equality (2.7). Denote the left-hand side of equality by I. Taking into account (2.6) we obtain

$$I \equiv \int_0^\infty \Delta_t\Big(\frac{\Psi_\sigma(w)}{\sqrt{u^2+\alpha^2}}\Big)\cos\lambda u\,du = 4s\int_0^\infty \frac{d^2}{ds^2}\Big[\frac{\Psi_\sigma(w)}{\sqrt{u^2+\alpha^2}}\Big]\cos\lambda u\,du$$
$$+\,4\int_0^\infty \frac{d}{ds}\Big[\frac{\Psi_\sigma(w)}{\sqrt{u^2+\alpha^2}}\Big]\cos\lambda u\,du + \int_0^\infty \frac{d^2}{dy_3^2}\Big[\frac{\Psi_\sigma(w)}{\sqrt{u^2+\alpha^2}}\Big]\cos\lambda u\,du.$$

Differentiating this equality we have

$$4\,\frac{d}{ds}\,\frac{\Psi_\sigma(w)}{\sqrt{u^2+\alpha^2}} = 4\Big[\frac{i\Psi_\sigma'(w)}{2(u^2+s)} - \frac{\Psi_\sigma(w)}{2(u^2+s)\sqrt{u^2+s}}\Big],$$

$$4s\,\frac{d^2}{ds^2}\,\frac{\Psi_\sigma(w)}{\sqrt{u^2+\alpha^2}} = 4s\Big[-\frac{\Psi_\sigma''}{4(u^2+s)^{3/2}} - \frac{i\Psi_\sigma'}{2(u^2+s)^2}$$
$$-\frac{i\Psi_\sigma'}{4(u^2+s)^2} + \frac{3\Psi_\sigma}{4(u^2+s)^2\sqrt{u^2+s}}\Big],$$

$$\frac{d^2}{dy_3^2}\Big[\frac{\Psi_\sigma(w)}{\sqrt{u^2+s^2}}\Big] = \frac{\Psi_\sigma''}{\sqrt{u^2+s^2}}.$$

Grouping the coefficients we shall obtain

$$I \equiv \int_0^\infty \frac{u^2\Psi_\sigma''\cos\lambda u}{(u^2+s)^{3/2}} + \int_0^\infty \frac{(2u^2-s)i}{(u^2+s)^2}\,\Psi'\cos\lambda u\,du$$
$$+\int_0^\infty \frac{s-2u^2}{(u^2+s)^2}\,\Psi_\sigma\cos\lambda u\,du = I_1+I_2+I_3.$$

Since

$$d_u\Psi_\sigma' = \frac{iu\Psi_\sigma''du}{\sqrt{u^2+s}},$$

then, integrating the first integral by parts, we have

$$I_1 = i\int_0^\infty \Big(\frac{s-u^2}{(u^2+s)^2}\cos\lambda u - \frac{u\lambda\sin\lambda u}{u^2+s}\Big)\Psi_\sigma' du.$$

Hence

$$I_1+I_2 = i\int_0^\infty \frac{u^2\Psi_\sigma'}{(u^2+s)^2}\cos\lambda u\,du - i\int_0^\infty \frac{\Psi_\sigma' u\lambda\sin\lambda u}{u^2+s}\,du.$$

Integrating by parts the both integrals, where

$$d_u\Psi_\sigma = \frac{iu}{\sqrt{u^2+s}}\,\Psi_\sigma'\,du.$$

Then, the first integral will be equal to

$$-\int_0^\infty \Big[\Big(\frac{1}{(u^2+s)^2} - \frac{3u^2}{(u^2+s)^{5/2}}\Big)\cos\lambda u - \frac{\lambda u \sin\lambda u}{(u^2+s)^{3/2}}\Big]\Psi_\sigma\, du,$$

and the second

$$\int_0^\infty \Big[\frac{-u\lambda\sin\lambda u}{(u^2+s)^{3/2}} + \frac{\lambda^2\cos\lambda u}{\sqrt{u^2+s}}\Big]\Psi_\sigma\, du.$$

Thus,

$$I_1 + I_2 = \lambda^2 \int_0^\infty \Psi_\sigma \frac{\cos\lambda u}{\sqrt{u^2+s}}\, du - \int_0^\infty \frac{s-2u^2}{(u^2+s)^{5/2}}\Psi_\sigma \cos\lambda u\, du.$$

Since I_3 is equal to the second integral, we obtain

$$I = \lambda^2 \int_0^\infty \Psi_\sigma(w)\frac{\cos\lambda u}{\sqrt{u^2+s}}\, du.$$

Thus function $\psi(\overline{w})$ is obtained from $\psi(w)$ by the change i by $-i$; therefore, the computations do not change and the second equality from (2.7) is established also. □

Lemma 2.2. *If $\lambda(y) \in C^2(\overline{D})$, then, for $y \neq x$ the following formula holds*

$$\Delta_y \Phi_\sigma(x,y) - \lambda^2 \Phi_\sigma(x,y) = T_\sigma(x,y), \tag{2.8}$$

where

$$2\pi^2 T_\sigma(x,y) = \int_0^\infty \Big[2\sum_{k=1}^{3}\frac{d\lambda}{dy_k}\frac{dF_\sigma}{dy_k}\Big] u\sin\lambda u\, du$$
$$- \int_0^\infty u^2 F_\sigma \Big(\sum_{k=1}^{3}\Big(\frac{d\lambda}{dy_k}\Big)^2\Big)\cos\lambda u\, du - \int_0^\infty F_\sigma u \sin\lambda u\, \Delta\lambda\, du.$$

Proof. Apply the Laplace operator with respect to the variable y in formula (2.2). The differentiation under the integral sign is valid since $y \neq x$ and $\sigma > 0$. As a result we obtain

$$2\pi^2 \Delta_y \Phi_\sigma(y,x) = \int_0^\infty \Delta_y(F_\sigma \cos\lambda u)du,$$

where

$$\Delta_y(F_\sigma \cos\lambda u) = \cos\lambda u \Delta_y F_\sigma + 2\sum_{n=1}^{3} \frac{dF_\sigma}{dy_k}\frac{d\cos\lambda u}{dy_k} + F_\sigma \Delta(\cos\lambda u),$$

$$\frac{d\cos\lambda u}{dy_k} = -u\sin\lambda u\,\frac{d\lambda}{dy_k},$$

$$\Delta(\cos\lambda u) = -u^2\cos\lambda u \sum_{k=1}^{3}\left(\frac{d\lambda}{dy_k}\right)^2 - u\sin\lambda u\,\Delta\lambda.$$

Now, equality (2.8) follows from (2.5). □

Corollary. *If $\lambda(y)$ is constant in $\overline{D}$, then the function $\Phi_\sigma(x,y)$ is the fundamental solution of equation (1.1) and $G_\sigma(x,y)$ from equality (2.4) is the regular solution of equation (1.1) in E^3 including the point $y = x$.*

Now, let $U(y)$ be a solution of equation (1.1) from the class $C^2(D)\bigcap C^1(\overline{D})$. Then, by the Green formula, for each $x \in D$ we shall have [2]

$$U(x) = \int_D U(y)T_\sigma(x,y)\,dy + \int_{\partial D}\left[\Phi_\sigma(x,y)\,\frac{dU}{dn}(y) - U(y)\,\frac{d\Phi_\sigma}{dn}(x,y)\right]ds_y,$$

where $T_\sigma(x,y)$ is defined in (2).

Denote

$$U_\sigma(x) = \int_S \left[\Phi_\sigma(x,y)g(y) - f(y)\,\frac{d\Phi_\sigma}{dn}(x,y)\right]ds_y, \quad x \in D, \tag{2.9}$$

$$R_\sigma(x) = \int_T \left[\Phi_\sigma(x,y)\,\frac{dU}{dn}(y) - U(y)\,\frac{d\Phi_\sigma}{dn}(x,y)\right]ds_y, \quad x \in D, \tag{2.10}$$

where $S \cup T = \partial D$. Then the formula for $U(x)$ will become as follows:

$$U(x) = \int_D U(y)T_\sigma(x,y)\,dy + U_\sigma(x) + R_\sigma(x) \qquad x \in D. \tag{2.11}$$

The functions $f(y)$ and $g(y)$ are known in S and belong to the class $C(S)$. Therefore, the function $U_\sigma(x)$ is defined everywhere except for the points from the surface S. As the difference of the simple and double layers the function $U_\sigma(x)$is extended continuously from D into $\overline{D}$. Since the desired solution *a priori* belongs to the class $C^2(D) \cap C^1(\overline{D})$, the function $R_\sigma(x)$ is defined via the values of solution and its normal derivative on the part T of the boundary ∂D which are not given *a priori*. We only assume that they are continuous and, therefore, bounded. However, as it will follow from further considerations, for large values of the parameter σ, the function $R_\sigma(x)$, $x \in D$ becomes small. Therefore, in the right-hand side of equality (2.11), we shall suppose that the term outside the integral $U_\sigma(x) + R_\sigma(x)$ is known. So, (2.11) can be considered as the Fredholm integral equation of the second kind with the kernel $T_\sigma(x, y)$ and with the unknown function $U(x)$.

For construction of solution of integral equation (2.11) we introduce the new functions

$$V(x) = e^{\sigma x_3^2} U(x), \qquad V_\sigma(x) = e^{\sigma x_3^2} U_\sigma(x),$$
$$P_\sigma(x) = e^{\sigma x_3^2} R_\sigma(x), \qquad K_\sigma(x, y) = e^{-\sigma(y_3^2 - x_3^2)} T_\sigma(x, y).$$

Then, relatively the new unknown function $V(x)$, equation (2.11) will become as follows:

$$V(x) = \int_D V(y) K_\sigma(x, y)\, dy + V_\sigma(x) + P_\sigma(x). \tag{2.12}$$

From equality (2) it follows that the function $T_\sigma(x, y)$for $y \neq x$ is continuous with respect to both variables x and y and is bounded in the neighborhood of the point $y = x$. Therefore, the solution of equation (2.12) can be constructed by the Fredholm method. So, for each $x \in D$ we shall have

$$V(x) = \int_D \Gamma_\sigma(x, y) \Big[V_\sigma(y) + P_\sigma(y) \Big] dy + V_\sigma(x) + P_\sigma(x), \tag{2.13}$$

where $\Gamma_\sigma(x, y)$ is the resolvent defined by the Newmann series

$$\Gamma_\sigma(x, y) = \sum_{n=1}^{\infty} K_\sigma^{(n)}(x, y). \tag{2.14}$$

The iterated kernels in this series are defined recurrently from the relations

$$K_\sigma^{(1)}(x, y) \equiv K_\sigma(x, y),$$
$$K_\sigma^{(n+1)}(x, y) = \int_D K_\sigma(x, t) K_\sigma^{(n)}(t, y)\, dt, \quad k = 1, 2, \ldots. \tag{2.15}$$

We show below that for large values of σ series (2.14) converges absolutely and uniformly in $\overline{D}$ relatively the variables x, y.

3. CONSTRUCTION OF SOLUTION AND THE REGULARIZATION BY LAVRENT'EV M. M.

Construction of the solution of problem (1.1)–(1.2) is based on the idea and method of Lavrent'ev M. M. [3, 4].

First we obtain the estimates for $K_\sigma(x,y)$, $\Gamma_\sigma(x,y)$, and $\Phi_\sigma(x,y)$.

Lemma 3.1. *For each $\sigma > 0$, $x, y \in \overline{D}$ the following equality*

$$2\pi^2 |K_\sigma(x,y)| \le [m + \psi(\sigma)] e^{-\sigma\alpha^2}, \tag{3.1}$$

holds, where

$$m = 6aa_5(\pi + 4\psi_0) + [(4+3\pi)(a_3+a_4) + (2+\pi)(a_3+a_4+a_5) \\ + (6+\pi)a_5]a_1 + 2[2(a_3+a_4) + 3a_5]a_2,$$

$$\psi(\sigma) = [\pi a a_5 + (\pi a_5 + 3\pi(a_3+a_4) + (a_3+a_4+a_5)(2+\pi))a_2 \\ + 2\pi a(a_3+a_4)]e^{a_2^2/(4\sigma)} + 3\sqrt{\pi} a_1 a_6/(4\sqrt{\sigma}) + [3(a_3^2+a_4^2+a_5^2+a_2a_6) \\ + 4a_2^2(2(a_3+a_4)+3a_5)]\sqrt{\pi}e^{a_2^2/(4\sigma)}/(4\sqrt{\sigma}) \\ + a_5[6a(a_1^2+a_2^2) + 96a_1^3 + 24a_1^2a_2 + 24a_1a_2^2 + 8a_2^3 \\ + (4a_2(6a_1^2a_2 + 3a_1a_2^2 + 2a_2^3) + 3aa_2^2)\sqrt{\pi}e^{a_2^2/(4\sigma)}/(2\sqrt{\sigma}) \\ + 72ae^{-\sigma}]/(12\sigma), \tag{3.2}$$

where $\psi(\sigma) \le \psi(1)$, $\sigma \ge 1$ and

$$a_1 = \max|\lambda_1(y)|; \quad a_2 = \max|\lambda_2(y)|; \quad a = \max_{\bar{D}}|\lambda|,$$

$$a_{k+2} = \max\left|\frac{d\lambda(y)}{dx_k}\right|, \quad k = 1,2,3; \qquad a_6 = \max|\Delta\lambda(y)|; \tag{3.3}$$

the maximum is taken along the closed domain $\overline{D}$,

$$\sup_{t\ge 0}\left|\int_0^t \frac{\sin\eta\, d\eta}{\eta}\right| = \psi_0. \tag{3.4}$$

Lemma 3.2. *Suppose the surface S is set by the equation*

$$y_3 = y_3(y_1, y_2), \qquad (y_1, y_2) \in \bar{T}, \tag{3.5}$$

where $y_3(y_1,y_2)$ is the univalent function satisfying the Lyapunov conditions. Denote

$$\max_{\bar{T}} y_3(y_1, y_2) = b. \tag{3.6}$$

Let $\sigma \ge 1$ and

$$\sigma > \pi\gamma^2 b|D|/2, \qquad 2\pi^2\gamma = m + \psi(1), \tag{3.7}$$

where $|D|$ is the volume of D; m and $\psi(1)$ are defined from (3.2). Then series (2.14) converges absolutely and uniformly for $x, y \in \overline{D}$. For the resolvent the following equalities hold

$$|\Gamma_\sigma(x,y)| \le \gamma e^{-\sigma\alpha^2} + \frac{\gamma^2}{\sqrt{2}} \frac{\sqrt{\pi b|D|}}{\sqrt{2\sigma} - \sqrt{\pi b \gamma^2 |D|}}, \tag{3.8}$$

$$\int_D |\Gamma_\sigma(x,y)|\, dy \le \frac{\gamma\sqrt{\pi b|D|}}{\sqrt{2\sigma}} \Big(1 + \frac{\gamma|D|\sqrt{\sigma}}{\sqrt{2\sigma} - \gamma\sqrt{\pi b|D|}}\Big). \tag{3.9}$$

Lemma 3.3. For each $\sigma > 0$, $x, y \in \overline{D}$ and $x \ne y$ the following equality holds

$$2\pi^2 e^{-\sigma(y_3^2 - x_3^2) + \sigma\alpha^2} |\Phi_\sigma(x,y)| \le \Big(\pi + 2\psi_0 + \frac{e^{-\sigma}}{2\sigma}\Big)\frac{1}{r} + \frac{3}{2} A_0(\sigma). \tag{3.10}$$

If $y_3 = 0$, then

$$2\pi^2 e^{\sigma(x_3^2 + \alpha^2)} |\Phi_\sigma(x,y)| \le \pi/2r + A_0(\sigma), \tag{3.11}$$

where

$$A_0(\sigma) = (2a_1^2 + a_2^2 + a_2^2 e^{a_2^2/(4\sigma)})\sqrt{\pi}/(8\sqrt{\sigma}), \tag{3.12}$$

$$2\pi^2 e^{\sigma\alpha^2 - \sigma(y_3^2 - x_3^2)} \Big|\frac{d\Phi_\sigma(x,y)}{dn}\Big| \le E(\sigma) + E_1(\sigma, x, y), \tag{3.13}$$

$$\begin{aligned} E(\sigma) = {} & \big[2(5+\pi)a_1^2 + (20+\pi)a_2^2 + (2+\pi)a_2^2 e^{a_2^2/(4\sigma)}\big]/16 \\ & + (5\pi + 12\psi_0 + 3e^{-\sigma}/\sigma)\sigma + (3\sqrt{\pi}/16)\big[12a_1(a_3 + a_4 + a_5) \\ & + a_2[12(a_3 + a_4 + a_5) + a_2^2] e^{a_2^2/(4\sigma)}\big]/\sqrt{\sigma}, \end{aligned} \tag{3.14}$$

$$\begin{aligned} E_1(\sigma, x, y) &= A(\sigma) y_3 + B(\sigma) y_3/r + C(\sigma)\alpha + D(\sigma)|\cos\Theta|/r^2, \\ A(\sigma) &= (3\sqrt{\pi}/16)\big(23a_1^2 + 28a_2^2 + 9a_2^2 e^{a_2^2/(4\sigma)}\big)/\sqrt{\sigma}, \\ B(\sigma) &= (3\pi + 2\psi_0 + e^{-\sigma}/(2\sigma))2\sigma, \\ C(\sigma) &= [3\sqrt{\pi}/(8\sqrt{\sigma})]\big(2a_1^2 + a_2^2 + a_2^2 e^{a_2^2/(4\sigma)}\big)\sqrt{\sigma}, \\ D(\sigma) &= 3\pi/2 + 4\psi_0 + e^{-\sigma}/\sigma, \end{aligned} \tag{3.15}$$

where θ is the angle between the radius vector $\overline{r}$, $x \in D$ and the normal in the point $y \in S$; a_k and u_0 are defined from (3.3) and (3.4) respectively.

If $y_3 = 0$, then

$$2\pi^2 e^{\sigma\alpha^2 + \sigma x_3^2} \Big|\frac{d\Phi_\sigma}{dy_3}(x,y)\Big| \le \frac{\pi x_3}{2r^3} + E_2(\sigma), \tag{3.16}$$

where

$$E_2(\sigma) = \big(4a_1^2 + 3a_2^2 + a_2^2 e^{a_2^2/(4\sigma)}\big)/8 \\ + (\sqrt{\pi}/16)\big[a_1 a_3 + a_2(a_2^2 + 8a_3)e^{a_2^2/(4\sigma)}\big]/\sqrt{\sigma} + \pi\sigma. \quad (3.17)$$

The proof of the Lemmas 3.1–3.3 can be found in Section 4.

Solution (2.13) of equation (2.12) we rewrite in the notations $U(x)$, $U_\sigma(x)$, $R_\sigma(x)$:

$$U(x) = H_\sigma(x) + h_\sigma(x), \quad H_\sigma(x) = U_\sigma(x) + e^{-\sigma x_3^2} \int_D \Gamma_\sigma(x,y) u_\sigma(y) e^{\sigma y_3^2}\, dy, \quad (3.18)$$

$$h_\sigma(x) = R_\sigma(x) + e^{-\sigma x_3^2} \int_D \Gamma_\sigma(x,y) R_\sigma(y) e^{\sigma y_3^2}\, dy, \quad (3.19)$$

where $U_\sigma(x)$, $R_\sigma(x)$ are defined by equalities (2.9), (2.10) respectively.

Further, we shall use the following elementary inequalities

$$\int_T e^{-\sigma\alpha^2}\, dy_1\, dy_2 \le \int_{-\infty}^{\infty}\int_{-\infty}^{\infty} e^{-\sigma\alpha^2}\, dy_1\, dy_2 \le \frac{\pi}{\sigma},$$

$$\int_T \alpha e^{-\sigma\alpha^2}\, dy_1\, dy_2 \le \frac{\pi\sqrt{\pi}}{2\sigma^{3/2}},$$

$$\int_T \frac{e^{-\sigma\alpha^2}}{r}\, dy_1\, dy_2 \le \frac{\pi\sqrt{\pi}}{\sqrt{\sigma}}, \qquad x_3 \int_T \frac{dy_1\, dy_2}{r^3} \le 2\pi. \quad (3.20)$$

Theorem 3.1. *Suppose that the surface S satisfies the conditions of Lemma 2.2 and σ ($\sigma \ge 1$) is chosen following condition (3.7). Suppose $U(x)$ is the solution of problem (1.1)–(1.2) from the class $C^1(\overline{D}) \cup C^2(D)$, where $f(y)$ and $g(y)$ set in S are the functions from the classes $C^1(S)$ and $C(S)$ respectively. Then for each $x \in D$ the following inequality holds*

$$|U(x) - H_\sigma(x)| \le M E_3(\sigma) e^{-\sigma x_3^2}, \qquad \lim_{\sigma\to\infty} E_3(\sigma) = 1, \quad (3.21)$$

where

$$E_3(\sigma) = \Big(\frac{1}{2} + \frac{\sqrt{\pi}}{4\sqrt{\sigma}} + \frac{A_0(\sigma) + E_2(\sigma)}{2\pi\sigma}\Big) \\ \times \Big(1 + \frac{\gamma\sqrt{\pi b|D|}}{\sqrt{2\sigma}}\Big(1 + \frac{|D|\gamma\sqrt{\sigma}}{\sqrt{2\sigma} - \gamma\sqrt{\pi b|D|}}\Big)\Big), \quad (3.22)$$

M, Γ, $A_0(\sigma)$, and $E_2(\sigma)$ are defined as in (1.3), (3.7), (3.12), and (3.17) respectively.

Proof. Taking into account (3.18), we have for $x \in D$

$$|U(x) - H_\sigma(x)| = |h_\sigma(x)|.$$

The theorem will follow from the equality

$$|h_\sigma(x)| \leq ME_3(\sigma)e^{-\sigma x_3^2}. \tag{3.23}$$

We shall establish it. Equality (3.19) yields

$$|h_\sigma(x)| \leq |R_\sigma(x)| + e^{-\sigma x_3^2} \int_D |\Gamma_\sigma(x,y)R_\sigma(y)|e^{\sigma y_3^2}\, dy. \tag{3.24}$$

For each $x \in D$, taking into account (1.3), (2.10), (3.20), (3.11), and (3.16), we obtain

$$\begin{aligned} |R_\sigma(x)| &\leq M \int_T \Big[|\Phi_\sigma| + \Big|\frac{d\Phi_\sigma}{dy_3}\Big|\Big]\, dy_1\, dy_2 \\ &\leq Me^{-\sigma x_3^2}\Big(\frac{1}{2} + \frac{\sqrt{\pi}}{4\sqrt{\sigma}} + \frac{A_0(\sigma) + E_2(\sigma)}{2\pi\sigma}\Big), \end{aligned}$$

where $A_0(\sigma)$ and $E_2(\sigma)$ are defined from (3.12) and (3.17) respectively. Finally, from (3.9) and (3.24), (3.23) follows. □

Corollary. *For each $x \in D$ the following limit holds:*

$$\lim_{\sigma \to \infty} U_\sigma(x) = U(x).$$

The limit holds uniformly in compact sets in D.

Now, we shall construct the regularization. The function $U_\sigma(x)$defined by equality (2.9) we represent as follows:

$$U_\sigma(x) = U_{\sigma\delta}(x) + I_{\sigma\delta}(x), \quad U_{\sigma\delta}(x) = \int_S \Big[g_\delta\Phi_\sigma - f_\delta\,\frac{d\Phi_\sigma}{dn}\Big]\, ds_y,$$

$$I_{\sigma\delta} = \int_S \Big[(g - g_\delta)\Phi_\sigma - (f - f_\delta)\frac{d\Phi_\sigma}{dn}\Big]\, ds_y, \tag{3.25}$$

where $f_\delta = f_\delta(y)$ and $g_\delta = g_\delta(y)$ are given function from the class $C(S)$ satisfying condition (1.4).

Denote

$$H_{\sigma\delta}(x) = U_{\sigma\delta} + e^{-\sigma x_3^2} \int_D \Gamma_\sigma(x,y)U_{\sigma\delta}(y)e^{\sigma y_3^2}\, dy, \tag{3.26}$$

$$G_{\sigma\delta}(x) = e^{-\sigma x_3^2} \int_D \Gamma_\sigma(x,y)I_{\sigma\delta}(y)e^{\sigma y_3^2}\, dy. \tag{3.27}$$

Then (3.18) can be rewritten as follows:

$$U(x) - H_{\sigma\delta}(x) = h_\sigma(x) + I_{\sigma\delta}(x) + G_{\sigma\delta}(x). \tag{3.28}$$

Further, we shall use the following elementary inequalities which follow easily from (3.20):

$$\int_S e^{-\sigma\alpha^2} ds_y \le \frac{c\pi}{\sigma}, \quad \int_S \frac{e^{-\sigma\alpha^2} ds_y}{r} \le \frac{c\pi\sqrt{\pi}}{\sqrt{\sigma}}, \quad \int_S \alpha e^{-\sigma\alpha^2} ds_y \le \frac{c\pi\sqrt{\pi}}{2\sigma\sqrt{\sigma}},$$

$$\int_S \frac{e^{-\sigma\alpha^2} y_3}{r} ds_y \le \frac{c\pi\sqrt{\pi}b}{\sqrt{\sigma}}, \qquad \int_S \frac{|\cos\Theta|}{r^2} ds_y \le 4\pi k,$$

$$c = \max_{\bar{T}} \sqrt{1 + \left(\frac{\partial y_3}{\partial y_1}\right)^2 + \left(\frac{\partial y_3}{\partial y_2}\right)^2}. \tag{3.29}$$

Here b is defined in (3.6), k denotes the maximal number of points intersection the surface δ with the vector r, where $x \in D$ and $y \in S$. Denote

$$\begin{aligned}
E_4(\sigma) &= (2k/\pi)(3\pi/2 + 4\psi_0 + e^{-\sigma}/\sigma) + [c/(2\pi)](5\pi + 12\psi_0 + 3e^{-\sigma}/\sigma) \\
&\quad + [c/(4\pi)]\{2\sqrt{\pi}b(3\pi + 2\psi_0 + e^{-\sigma}/(2\sigma))\sqrt{\sigma} \\
&\quad + (\sqrt{\pi}/16)\left[b(23a_1^2 + 28a_2^2 + 9a_2^2 e^{a_2^2/(4\sigma)}) + 16(\pi + 2\psi_0 + e^{-\sigma}/\sigma)/\sqrt{\sigma}\right] \\
&\quad + (1/8)\left[(5 + 4\pi)a_1^2 + 2(5 + \pi)a_2^2 + (1 + 2\pi)a_2^2 e^{a_2^2/(4\sigma)}\right]/\sigma \\
&\quad + (3\sqrt{\pi}/16)\left[2a_1^2 + a_2^2 + 4a_1(a_3 + a_4 + a_5)\right. \\
&\quad \left. + a_2[4(a_3 + a_4 + a_5) + a_2 + a_2^2]e^{a_2^2/(4\sigma)}\right]/(\sigma\sqrt{\sigma})\},
\end{aligned} \tag{3.30}$$

where a_k and ψ_0 are defined from (3.3) and (3.4) respectively.

Theorem 3.2. *Suppose that the surface S satisfies the conditions of Lemma 3.2 and σ ($\sigma \ge 1$) is chosen following condition (3.7). Suppose also that $U(x)$ solves problem (1.1)–(1.2), belongs to the class $C^2(D) \bigcap C^1(\overline{D})$ where $f(y) \in C^1(S)$, $g_y \in C(S)$. Let, instead of $f(y)$ and $g(y)$, their continuous approximations $f_\delta(y)$, $g_\delta(y)$ respectively be given satisfying condition (1.4); besides, condition (1.3) holds, where b and M are given positive numbers so that*

$$\delta < Me^{-\pi\gamma^2 b^3 |D|/2}. \tag{3.31}$$

Then for each $x \in D$ the inequality

$$|U(x) - H_{\sigma\delta}(x)| \le T(\sigma) M^{1 - x_3^2/b^2} \delta^{x_3^2/b^2}, \tag{3.32}$$

holds, where

$$\sigma = \frac{1}{b^2} \ln \frac{M}{\delta}, \tag{3.33}$$

$$T(\sigma) = \Big[\frac{1}{2} + \frac{\sqrt{\pi}}{4\sqrt{\sigma}} + E_4(\sigma) + \frac{A_0(\sigma) + E_2(\sigma)}{2\pi\sigma}\Big] \times \Big[1 + \frac{\gamma\sqrt{\pi b|D|}}{\sqrt{2\sigma}}\Big(1 + \frac{|D|\gamma\sqrt{\sigma}}{\sqrt{2\sigma} - \gamma\sqrt{\pi b|D|}}\Big)\Big] \tag{3.34}$$

and $A_0(\sigma)$, $E_2(\sigma)$, $E_4(\sigma)$ *are defined in* (3.12), (3.17), *and* (3.30) *respectively;* $T(\sigma) = O(\sqrt{\sigma})$, $\sigma \to \infty$.

Proof. For each $x \in D$ and σ ($\sigma \geq 1$) satisfying condition (3.7), the following inequalities hold

$$|I_{\sigma\delta}(x)| \leq E_4(\sigma) e^{\sigma b^2 - \sigma x_3^2}\,\delta, \tag{3.35}$$

$$|G_{\sigma\delta}(x)| \leq E_5(\sigma) e^{\sigma b^2 - \sigma x_3^2}\,\delta, \tag{3.36}$$

where

$$E_5(\sigma) = \frac{\gamma\sqrt{\pi b|D|}}{\sqrt{2\sigma}}\Big(1 + \frac{|D|\gamma\sqrt{\sigma}}{\sqrt{2\sigma} - \gamma\sqrt{\pi b|D|}}\Big) E_4(\sigma), \tag{3.37}$$

E_4 is defined by equality (3.30).

First, we prove (3.35). For each $x \in D$, (3.25) and (1.4) yields

$$|I_{\sigma\delta}| \leq \delta\Big[\int_S |\Phi_\sigma| ds_y + \int_S \Big|\frac{d\Phi_\sigma}{dn}\Big| ds_y\Big].$$

The moduli $|\Phi_\sigma|$ and $|\,d\Phi_\sigma/\,dn|$ in the right-hand side we replace by estimate (3.10) and (3.13). The elementary integrals we replace by estimates (3.29). The obtained expression we group by decreasing degrees of $\sqrt{\sigma}$. As a result we obtain (3.35). Further, (3.27) and (3.35) yield the estimate

$$|G_{\sigma\delta}| \leq E_4(\sigma) e^{\sigma b^2 - \sigma x_3^2}\,\delta \int_D |\Gamma_\sigma(x, y)|\, dy.$$

Now, (3.36) follows from (3.9).

Formulas (3.36), (3.35), (3.23), and (3.28) yield the following inequality for arbitrary $x \in D$.

$$|U(x) - H_{\sigma\delta}(x)| \leq \big[M E_3(\sigma) + \delta(E_4(\sigma) + E_5(x)) e^{b^2\sigma}\big] e^{-\sigma x_3^2}.$$

Choose σ which satisfies (3.33). As a result we obtain (3.32). □

Corollary. *For each $x \in D$ the limit equality holds*

$$\lim_{\delta \to 0} H_{\sigma\delta}(x) = U(x).$$

The limit attains uniformly in compact sets from D.

The linear functional $H_{\sigma\delta}(x) = H_{\sigma\delta}(f_\delta, g_\delta, x)$ defined in the set of pairs (f_δ, g_δ), where f_δ, $g_\delta(y) \in C(S)$ and $0 \le \delta \le \delta_0$ is called *the regularization following Lavrent'ev, M. M.* of solution of problem (1.1)–(1.2).

4. THE PROOF OF LEMMAS 3.1–3.3

The proof of Lemma 3.1 we divide into four lemmas.

From (2) for the function $K_\sigma(x, y)$ we have

$$2\pi^2 e^{\sigma\alpha^2} K_\sigma(x,y) = 2\pi^2 e^{-\sigma(y_3^2 - x_3^2)+\sigma\alpha^2} T_\sigma(x,y) = T_1 + T_2 + T_3 + T_4 + T_5, \quad (4.1)$$

where

$$T_1 = 2\sum_{k=1}^{3} \frac{d\lambda}{dy_k} \frac{d}{dy_k}\left(\sigma(y_3^2 - x_3^2) - \sigma\alpha^2\right) \int_0^\infty \frac{u \sin \lambda u \varphi_\sigma}{u^2 + r^2} e^{-\sigma u^2}\, du,$$

$$T_2 = 2\sum \frac{d\lambda}{dy_k} \int_0^\infty u \sin \lambda u e^{-\sigma u^2} \frac{d\varphi_\sigma}{dy_k} \frac{du}{u^2 + r^2},$$

$$T_3 = -2\sum \frac{d\lambda}{dy_k} \int_0^\infty u \sin \lambda u e^{-\sigma u^2} \varphi_\sigma \frac{2(y_k - x_k)du}{(u^2 + r^2)^2},$$

$$T_4 = -\sum \left(\frac{d\lambda}{dy_k}\right)^2 \int_0^\infty u^2 \cos \lambda u e^{-\sigma u^2} \varphi_\sigma \frac{du}{u^2 + r^2},$$

$$T_5 = -\Delta\lambda \int_0^\infty u \sin \lambda u e^{-\sigma u^2} \varphi_\sigma \frac{du}{u^2 + r^2}.$$

Here $\varphi_\sigma = \varphi_\sigma(x, y, u)$ is defined by formula (2.3).

Lemma 4.1. *For each $\sigma > 0$, $x, y \in \overline{D}$ the following inequality holds*

$$|T_4 + T_5| \le [3\sqrt{\pi}/(4\sqrt{\sigma})]\left[a_1 a_6 + (a_3^2 + a_4^2 + a_5^2 + a_2 a_6)e^{a_2^2/(4\sigma)}\right], \quad (4.2)$$

where a_k, $k = \overline{1,6}$ are defined in (3.3).

Proof. Taking into account (2.3), (3.3), inequality $|\cos h| \le \operatorname{ch} \lambda_2$ and equality

$$\int_0^\infty ch\lambda_2 u e^{-\sigma u^2}\, du = \frac{\sqrt{\pi}}{2\sqrt{\sigma}} e^{\lambda_2^2/(4\sigma)},$$

we obtain

$$|T_4| \le (a_3^2 + a_4^2 + a_5^2)[3\sqrt{\pi}/(4\sqrt{\sigma})]e^{a_2^2/(4\sigma)}.$$

Similarly, taking into account the inequalities

$$|\sin \lambda u| \le |\sin \lambda_1 u| + sh\lambda_2 u, \qquad |\sin \lambda_1 u| \le |\lambda_1| u$$

we obtain

$$|T_5| \le [3\sqrt{\pi}/(4\sqrt{\sigma})]a_6\big(a_1 + a_2 e^{a_2^2/(4\sigma)}\big).$$

Summing these estimates for T_4 and T_5 we obtain (4.2). □

Lemma 4.2. *For each* $\sigma > 0$, $x, y \in \overline{D}$, *the inequality*

$$|T_3| \le (\pi + 2)(a_3 + a_4 + a_5)\big(a_1 + a_2 e^{a_2^2/(4\sigma)}\big) \tag{4.3}$$

holds, where a_k, $k = \overline{1,5}$ *are defined in* (3.3).

Proof. Analogously Lemma 4.1 we have

$$|T_3| \le 2\sum_{k=1}^{3} a_{k+2}\Big[\int_0^\infty (a_1 u + a_2 u \operatorname{ch} a_2 u)\frac{2ur}{(u^2+r^2)^2}\, e^{-\sigma u^2}\, du + \int_0^\infty (a_1 u + a_2 u \operatorname{ch} a_2 u)\frac{2r^2}{(u^2+r^2)^2}\, e^{-\sigma u^2}\, du\Big]$$
$$\le 2\sum_{k=1}^{3} a_{k+2}\Big[2\Big(a_1 \frac{\pi}{4} + a_2 \frac{\pi}{4} e^{a_2^2/(4\sigma)}\Big) + a_1 + a_2 e^{a_2^2/(4\sigma)}\Big].$$

Hence (4.3) follows. □

Lemma 4.3. *For each* $\sigma > 0$; $x, y \in \overline{D}$, *the inequality*

$$|T_2| \le (a_3 + a_4)[(2+3\pi)a_1 + 2a_2] + a_5(2a(\pi + 4\psi_0) + 3(a_1 + a_2)) + \psi_2(\sigma) \tag{4.4}$$

holds, where

$$\psi_2(\sigma) = \pi(a_3 + a_4)(2a + 3a_2)e^{a_2^2/(4\sigma)} + [2(a_3 + a_4) + 3a_5][\sqrt{\pi}a_2^2/(2\sqrt{\sigma})]e^{a_2^2/(4\sigma)} + 2a_5(\psi_1(\sigma) + ae^{-\sigma}/\sigma), \tag{4.5}$$

$$\psi_1(\sigma) = [24a_1^3 + 6a_1a_2^2 + 2a_2^3 + 6a_1^2a_2 + a_2(3a_1a_2^2 + 2a_2^3 + 6a_1^2a_2)[\sqrt{\pi}/(2\sqrt{\sigma})]e^{a_2^2/(4\sigma)}]/(24\sigma). \tag{4.6}$$

Proof. We represent T_2 as follows:

$$T_2 = 2\sum_{k=1}^{2} \frac{d\lambda_k}{dy_k} I_k + \frac{d\lambda_3}{dy_3} I_3, \tag{4.7}$$

where

$$I_k = \int_0^\infty \frac{d\varphi_k}{dy_k} u \sin \lambda u \, \frac{e^{-\sigma u^2} du}{u^2 + r^2} \tag{4.8}$$

and

$$\frac{d\varphi_\sigma}{dy_k} = -\tau \frac{\sin \tau\sqrt{u^2+\alpha^2}}{\sqrt{u^2+\alpha^2}}(y_k - x_k) - \tau(y_3 - x_3)\frac{\cos \tau\sqrt{u^2+\alpha^2}}{\sqrt{u^2+\alpha^2}} \frac{y_k - x_k}{\sqrt{u^2+\alpha^2}}$$
$$+ (y_3 - x_3)\frac{y_k - x_k}{\sqrt{u^2+\alpha^2}} \frac{\sin \tau\sqrt{u^2+\alpha^2}}{u^2+\alpha^2}, \quad k = 1, 2, \tag{4.9}$$

$$\frac{d\varphi_\sigma}{dy_3} = -2\sigma \sin \tau\sqrt{u^2+\alpha^2}\sqrt{u^2+\alpha^2} - \frac{\sin \tau\sqrt{u^2+\alpha^2}}{\sqrt{u^2+\alpha^2}}$$
$$- (y_3 - x_3)2\sigma \cos \tau\sqrt{u^2+\alpha^2}.$$

We show that

$$|I_3| \le \frac{2a(\pi + 4\psi_0) + 3(a_1 + a_2)}{2} + \frac{3}{4}\frac{\sqrt{\pi}}{\sqrt{\sigma}} a_2^2 e^{a_2^2/(4\sigma)} + \psi_1(\sigma) + \frac{ae^{-\sigma}}{\sigma}, \tag{4.10}$$

Taking into account (4.8), (4.9), analogous to the proof of Lemmas 4.1 and 4.2, we deduce the inequality:

$$|I_3| \le 3\int_0^\infty \sigma(a_1 u + a_2 u \operatorname{ch} a_2 u)e^{-\sigma u^2} du + |B|$$
$$\le \frac{3}{2}\left(a_1 + a_2 + \frac{\sqrt{\pi}}{2\sqrt{\sigma}} a_2^2 e^{a_2^2/(4\sigma)}\right) + |B|, \tag{4.11}$$

where

$$B = -\int_0^\infty \frac{u \sin \lambda u}{u^2 + r^2} \frac{\sin \tau\sqrt{u^2+\alpha^2}}{\sqrt{u^2+\alpha^2}} e^{-\sigma u^2} du.$$

The evident representation

$$\sin \lambda u = \lambda u + f_u, \tag{4.12}$$

$$f_u = \lambda_1 u[\sin(\lambda_1 u)/(\lambda_1 u) - 1] - 2\sin\lambda_1 u\,\mathrm{sh}^2(\lambda_2 u/2) + i\{\lambda_2 u[\mathrm{sh}(\lambda_2 u)/(\lambda_2 u) - 1] - 2\sin^2(\lambda_1 u/2)\,\mathrm{sh}\,\lambda_2 u\}$$

and the inequality

$$|f_u| \le u^3[24|\lambda_1|^3 + 3|\lambda_1|\lambda_2^2 + (3|\lambda_1|\lambda_2^2 + 2|\lambda_2|^3 + 6\lambda_1^2|\lambda_2|)\,\mathrm{ch}\,|\lambda_2|u]/12 \quad (4.13)$$

yield

$$B = B_1 + B_2,$$

where

$$B_1 = -\int_0^\infty \frac{\lambda u^2 \sin\tau\sqrt{u^2+\alpha^2}}{\sqrt{u^2+\alpha^2}} \frac{e^{-\sigma u^2}\,du}{u^2+r^2},$$

$$B_2 = -\int_0^\infty \frac{u \sin\tau\sqrt{u^2+\alpha^2}}{\sqrt{u^2+\alpha^2}} \frac{f_u e^{-\sigma u^2}\,du}{u^2+r^2}.$$

Analogously the proofs of Lemmas 4.1 and 4.3, from (4.13) we deduce the inequality

$$|B_2| \le [24a_1^3 + 6a_1^2 a_2 + 6a_1 a_2^2 + 2a_2^3 + a_2(6a_1^2 a_2 + 3a_1 a_2^2 + 2a_2^3)[\sqrt{\pi}/(2\sqrt{\sigma})]e^{a_2^2/(4\sigma)}]/(24\sigma) \equiv \psi_1(\sigma).$$

Further we represent B_1 as the sum

$$B_1 = B_3 + B_4,$$

where

$$B_3 = -\lambda\int_0^\infty \frac{\sin\tau\sqrt{u^2+\alpha^2}}{\sqrt{u^2+\alpha^2}} e^{-\sigma u^2}\,du,$$

$$B_4 = \lambda r^2 \int_0^\infty \frac{\sin\tau\sqrt{u^2+\alpha^2}}{\sqrt{u^2+\alpha^2}} \frac{e^{-\sigma u^2}}{u^2+r^2}\,du, \quad \tau = 2\sigma y_3.$$

Estimate B_3. We have

$$-\lambda^{-1}B_3 = \int_0^{\sqrt{1+\alpha^2\tau^2}} \frac{\sin\tau\sqrt{u^2+\alpha^2}}{\sqrt{u^2+\alpha^2}} e^{-\sigma u^2}\,du + \int_{\sqrt{1+\alpha^2\tau^2}}^\infty \frac{\sin\tau\sqrt{u^2+\alpha^2}}{\sqrt{u^2+\alpha^2}} e^{-\sigma u^2}\,du.$$

The second integral we can estimate as follows:

$$\Big|\int_{\sqrt{1+\alpha^2\tau^2}}^{\infty}\frac{\sin\tau\sqrt{u^2+\alpha^2}}{\sqrt{u^2+\alpha^2}}\,e^{-\sigma u^2}\,du\Big|\le\int_1^{\infty}ue^{-\sigma u^2}du=\frac{1}{2\sigma}\,e^{-\sigma}.$$

The first integral by the Bonne theorem ($\psi(u)=e^{-\sigma u^2}$ is monotonically decreasing, $\psi(0)=1$) is equal to

$$\int_0^{\xi}\frac{\sin\tau\sqrt{u^2+\alpha^2}}{\sqrt{u^2+\alpha^2}}\,du=\int_{\alpha\tau}^{\tau\sqrt{\xi^2+\alpha^2}}\frac{\sin t\,dt}{\sqrt{t^2-\alpha^2\sigma^2}}=\int_{\alpha\tau}^{\tau\sqrt{\xi^2+\alpha^2}}\frac{\sin t\,dt}{t}$$
$$+\int_{\alpha\tau}^{\tau\sqrt{\xi^2+\alpha^2}}\frac{\alpha^2\tau^2}{t\sqrt{t^2-\alpha^2\tau^2}}\,\frac{\sin t\,dt}{t+\sqrt{t^2-\alpha^2\tau^2}},\quad 0<\xi\le\sqrt{1+\tau^2\alpha^2}.$$

Here the modulus of the first integral, following (3.4), is not greater than $2\psi_0$ and

$$\Big|\int_{\alpha\tau}^{\tau\sqrt{\xi^2+\alpha^2}}\frac{\alpha^2\tau^2}{t\sqrt{t^2-\alpha^2\tau^2}}\,\frac{\sin t\,dt}{t+\sqrt{t^2-\alpha^2\tau^2}}\Big|\le\int_1^{\infty}\frac{dt}{t^2\sqrt{t^2-1}}$$
$$=\int_0^{\infty}\frac{du}{(1+u^2)\sqrt{1+u^2}}<\frac{\pi}{2}.$$

Combining these inequalities, we obtain

$$|B_3|\le(\pi/2+2\psi_0+e^{-\sigma}/(2\sigma))|\lambda|. \tag{4.14}$$

Since the function $\psi(u)=(u^2+r^2)^{-1}e^{-\sigma u^2}$ monotonically decreases for $u\ge 0$ and $\psi(0)=r^{-2}$, then, this estimate holds for $|B_4|$ also. Therefore,

$$|B_1|\le(\pi+4\psi_0+e^{-\sigma}/\sigma)|\lambda|.$$

Combining the estimate for $|B_1|$ and $|B_2|$ and taking into account (4.11), we obtain (4.10).

We prove now the inequalities

$$|I_k|\le\frac{1}{2}\Big[(2+3\pi)a_1+2a_2+\pi(2a+3a_2)e^{a_2^2/(4\sigma)}+\frac{\sqrt{\pi}}{\sqrt{\sigma}}\,a_2^2e^{a_2^2/(4\sigma)}\Big],\quad k=1,2. \tag{4.15}$$

From (4.8), (4.9), for I_k, we have

$$I_k=B_{k1}+B_{k2}+B_{k3},$$

where

$$B_{k1} = -\tau(y_k - x_k)\int_0^\infty \frac{\sin\tau\sqrt{u^2+\alpha^2}}{\sqrt{u^2+\alpha^2}}\,\frac{u\sin\lambda u e^{-\sigma u^2}\,du}{u^2+r^2},$$

$$B_{k2} = -\tau(y_3 - x_3)(y_k - x_k)\int_0^\infty \frac{\cos\tau\sqrt{u^2+\alpha^2}}{\sqrt{u^2+\alpha^2}}\,\frac{u\sin\lambda u e^{-\sigma u^2}\,du}{\sqrt{u^2+\alpha^2}(u^2+r^2)},$$

$$B_{k3} = -(y_3 - x_3)(y_k - x_k)\int_0^\infty \frac{u\sin\tau\sqrt{u^2+\alpha^2}}{\sqrt{u^2+\alpha^2}(u^2+\alpha^2)}\,\frac{\sin\lambda u e^{-\sigma u^2}\,du}{u^2+r^2}.$$

Similarly to Lemma 4.1, we obtain the inequality

$$|B_{k3}| \le \frac{|y_3 - x_3|}{2}\Big(|\lambda_1|\int_0^\infty \frac{du}{u^2+r^2} + |\lambda_2|e^{\lambda_2^2/(4\sigma)}\int_0^\infty \frac{du}{u^2+r^2}\Big)$$
$$\le \pi(a_1 + a_2 e^{a_2^2/(4\sigma)})/4.$$

As

$$-\tau u\cos\tau\sqrt{u^2+\alpha^2}/\sqrt{u^2+\alpha^2} = d_u \sin\tau\sqrt{u^2+\alpha^2},$$

then, integrating by parts, we obtain

$$B_{k2} = -(y_3 - x_3)(y_k - x_k)\Big[\int_0^\infty \frac{\lambda e^{-\sigma u^2}\cos\lambda u\sin\tau\sqrt{u^2+\alpha^2}}{\sqrt{u^2+\alpha^2}(u^2+r^2)}\,du$$
$$+\int_0^\infty \frac{(-2u\sigma)e^{-\sigma u^2}\sin\lambda u\sin\tau\sqrt{u^2+\alpha^2}}{\sqrt{u^2+\alpha^2}(u^2+r^2)}\,du$$
$$+\int_0^\infty \frac{\sin\lambda u e^{-\sigma u^2}\sin\tau\sqrt{u^2+\alpha^2}}{\sqrt{u^2+\alpha^2}}\,\frac{-2u\,du}{(u^2+r^2)^2}$$
$$+\int_0^\infty \frac{u\sin\tau\sqrt{u^2+\alpha^2}}{\sqrt{u^2+\alpha^2}}\,\frac{e^{-\sigma u^2}\sin\lambda u}{(u^2+\alpha^2)(u^2+r^2)}\Big].$$

The integrals standing in the right-hand side we estimate as in Lemmas 4.1, 4.2. As a result we obtain

$$|B_{k2}| \le \frac{\pi}{2}|\lambda|e^{\lambda_2^2/(4\sigma)} + \int_0^\infty |\lambda_1|\sigma u e^{-\sigma u^2}du + \int_0^\infty \sigma|\lambda_2|u\operatorname{ch}\lambda_2 u e^{-\sigma u^2}$$
$$+r\Big(\int_0^\infty \frac{2|\lambda_1|u^2\,du}{(u^2+r^2)^2} + e^{\lambda_2^2/(4\sigma)}\int_0^\infty \frac{2|\lambda_2|u^2\,du}{(u^2+r^2)^2}\Big)$$
$$+\frac{r}{2}\Big(\int_0^\infty \frac{|\lambda_1|\,du}{u^2+r^2} + |\lambda_2|e^{\lambda_2^2/(4\sigma)}\int_0^\infty \frac{du}{u^2+r^2}\Big)$$
$$\le \frac{\pi}{2}ae^{a_2^2/(4\sigma)} + \frac{a_1}{2} + \frac{a_2}{2}\Big(1 + \frac{\sqrt{\pi}}{2\sqrt{\sigma}}a_2 e^{a_2^2/(4\sigma)}\Big) + \frac{\pi a_1}{2}$$
$$+\frac{\pi}{2}a_2 e^{a_2^2/(4\sigma)} + \frac{\pi}{4}\Big(a_1 + a_2 e^{a_2^2/(4\sigma)}\Big).$$

For B_{k1}, we have similarly

$$|B_{k1}| \le \frac{\pi}{2}\, a e^{a_2^2/(4\sigma)} + \frac{a_1}{2} + \frac{a_2}{2}\Big(1 + \frac{\sqrt{\pi}}{2\sqrt{\sigma}}\, a_2 e^{a_2^2/(4\sigma)}\Big) + \frac{\pi}{2}\Big(a_1 + a_2 e^{a_2^2/(4\sigma)}\Big).$$

Adding the inequalities for $|B_{k1}|$, $|B_{k2}|$, and $|B_{k3}|$ we obtain inequalities (4.15). □

Lemma 4.4. *For each $\sigma > 0$, $x, y \in \overline{D}$ the following inequality holds*

$$|T_1| \le 2(a_1+a_2)(a_3+a_4)+a_5[4a(\pi+4\psi_0)+(3+\pi)a_1+3a_2]+\psi_3(\sigma), \quad (4.16)$$

where

$$\begin{aligned}\psi_3(\sigma) &= \pi a_5(a+a_2)e^{a_2^2/(4\sigma)} + a_2^2(2a_3+2a_4+3a_5)[\sqrt{\pi}/(2\sqrt{\sigma})]e^{a_2^2/(4\sigma)} \\ &\quad + [a_5/(4\sigma)]\{2a(a_1^2+a_2^2) + 24a_1^3 + 6a_1^2a_2 + 6a_1a_2^2 + 2a_2^3 \\ &\quad + a_2(6a_1^2a_2 + 3a_1a_2^2 + 2a_2^3 + aa_2^2)[\sqrt{\pi}/(2\sqrt{\sigma})]e^{a_2^2/(4\sigma)} + 16ae^{-\sigma}\}.\end{aligned}$$

The numbers a, a_k, ψ_0 are defined in (3.3) *and* (3.4).

Proof. Represent T_1 in the form

$$T_1 = 2\Big(\frac{d\lambda}{dy_1}\, A_1 + \frac{d\lambda}{dy_2}\, A_2 + \frac{d\lambda}{dy_3}\, A_3\Big), \qquad (4.17)$$

where

$$A_k = -2\sigma(y_k - x_k)\int_0^\infty u \sin\lambda u e^{-\sigma u^2}\varphi_\sigma \frac{du}{u^2+r^2}, \qquad k = 1, 2$$

$$A_3 = \tau \int_0^\infty u \sin\lambda u e^{-\sigma u^2}\varphi_\sigma \frac{du}{u^2+r^2}, \qquad \tau = 2\sigma y_3.$$

Since $2u|y_k - x_k| \le u^2 + r^2$, analogously Lemma 2.1, we obtain

$$|A_k| \le a_1 + a_2 + [\sqrt{\pi}/(2\sqrt{\sigma})]a_2^2 e^{a_2^2/(4\sigma)}, \qquad k = 1, 2. \qquad (4.18)$$

Represent A_3 in the form

$$A_3 = A_4 + A_5, \qquad (4.19)$$

where

$$A_4 = \tau \int_0^\infty \cos\tau\sqrt{u^2+\alpha^2}\, u \sin\lambda u e^{-\sigma u^2}\, \frac{du}{u^2+r^2},$$

$$A_5 = \tau(y_3 - x_3)\int_0^\infty \frac{\sin\tau\sqrt{u^2+\alpha^2}}{\sqrt{u^2+\alpha^2}}\, u\sin\lambda u e^{-\sigma u^2}\, \frac{du}{u^2+r^2}. \qquad (4.20)$$

Since

$$\frac{\tau \sin \tau\sqrt{u^2+\alpha^2}}{\sqrt{u^2+\alpha^2}}\, u\,du = -d_u \cos \tau\sqrt{u^2+\alpha^2},$$

$$\frac{d}{du}\Big(\frac{\sin \lambda u e^{-\sigma u^2}}{u^2+r^2}\Big) = \frac{\lambda \cos \lambda u e^{-\sigma u^2}}{u^2+r^2} - \frac{2u\sigma \sin \lambda u e^{-\sigma u^2}}{u^2+r^2} - \frac{2u \sin \lambda u e^{-\sigma u^2}}{(u^2+r^2)^2},$$

then, integrating by parts A_5, we represent it as the sum of three integrals. Estimating the integrals as in Lemma 4.1, we obtain

$$|A_5| \le |\lambda| r \int_0^\infty \frac{\operatorname{ch} \lambda_2 u e^{-\sigma u^2}}{u^2+r^2}\,du + \sigma \int_0^\infty (|\lambda_1|u + |\lambda_2| u \operatorname{ch} \lambda_2 u) e^{-\sigma u^2}\,du$$
$$+ r \int_0^\infty 2u(|\lambda_1|u + |\lambda_2| u \operatorname{ch} \lambda_2 u) e^{-\sigma u^2} \frac{du}{(u^2+r^2)^2}.$$

Hence, for $|A_5|$ we obtain the final inequality

$$|A_5| \le \frac{(1+\pi)a_1 + a_2}{2} + \frac{\pi}{2}(a + a_2) e^{a_2^2/(4\sigma)} + \frac{\sqrt{\pi}}{4\sqrt{\sigma}}\, a_2^2 e^{a_2^2/(4\sigma)}. \qquad (4.21)$$

Integrating by parts in (4.20) and using the representation (4.12) and the formula $\cos n = 1 - 2\sin^2(u/2)$ we represent A_4 as the sum $A_4 = A_6 + A_7$, where

$$A_6 = \int_0^\infty 2u\sigma\sqrt{u^2+\alpha^2} \sin \tau\sqrt{u^2+\alpha^2}\, \frac{\sin \lambda u e^{-\sigma u^2} du}{u^2+r^2},$$

$$A_7 = \lambda \int_0^\infty \Big(-\frac{r^2+\beta^2}{u^2+r^2} + \frac{2r^2\beta^2}{(u^2+r^2)^2}\Big) e^{-\sigma u^2} \frac{\sin \tau\sqrt{u^2+\alpha^2}}{\sqrt{u^2+\alpha^2}}\, du$$
$$+ \int_0^\infty \Big(\frac{2u(u^2+\alpha^2)}{(u^2+r^2)^2} - \frac{u}{u^2+r^2}\Big) f_u e^{-\sigma u^2} \frac{\sin \tau\sqrt{u^2+\alpha^2}}{\sqrt{u^+\alpha^2}}\, du$$
$$+ 2\lambda \int_0^\infty \frac{u^2+\alpha^2}{u^2+r^2} \sin^2 \frac{\lambda u}{2}\, e^{-\sigma u^2} \frac{\sin \tau\sqrt{u^+\alpha^2}}{\sqrt{u^2+\alpha^2}}\, du$$

and

$$\frac{2u^2(u^2+\alpha^2)}{(u^2+r^2)^2} - \frac{u^2}{u^2+r^2} - \frac{u^2+\alpha^2}{u^2+r^2} = \frac{2\beta^2 r^2}{(u^2+r^2)^2} - \frac{r^2+\beta^2}{u^2+r^2}, \quad \beta = y_3 - x_3.$$

Estimate A_7. Similar to Lemma 4.3, the modulus of the sum of the first two terms in the right-hand side is not greater than

$$4a(\pi/2 + 2\psi_0 + \exp^{-\sigma}/2\sigma) + 3\psi_1(\sigma).$$

The modulus of the third term is not greater than

$$[a/(8\sigma)]\{2a_1^2 + 2a_2^2 + [\sqrt{\pi}/(2\sqrt{\sigma})]a_2^3 e^{a_2^2/(4\sigma)}\},$$

since

$$|\sin^2(\lambda u/2)| \le (u^2/8)(2a_1^2 + a_2^2 + a_2^2 \operatorname{ch} a_2 a_2).$$

Therefore, we obtain

$$|A_7| \le 2\Big(\pi + 4\psi_0 + \frac{e^{-\sigma}}{\sigma}\Big)a + \frac{a}{8\sigma}\Big[2a_1^2 + 2a_2^2 + \frac{\sqrt{\pi}}{2\sqrt{\sigma}}\, a_2^3 e^{a_2^2/(4\sigma)}\Big] + 3\psi_1(\sigma).$$

Further, since $wu\sqrt{u^2+\alpha^2} \le u^2 + r^2$, then $|A_6|$ satisfies inequality (4.18).

Therefore, for A_4 we obtain the estimate

$$|A_4| \le 2a\Big(\pi + 4\psi_0 + \frac{e^{-\sigma}}{\sigma}\Big) + \frac{a}{8\sigma}\Big[2a_1^2 + 2a_2^2 + \frac{\sqrt{\pi}}{2\sqrt{\sigma}}\, a_2^3\, e^{a_2^2/(4\sigma)}\Big]$$
$$+ 3\psi_1(\sigma) + a_1 + a_2 + \frac{\sqrt{\pi}}{2\sqrt{\sigma}} a_2^2 e^{a_2^2/(4\sigma)}.$$

Combining (4), (4.21), (4.19), (4.18), and (4.17) we obtain (4.16). □

Add the inequalities (4.16), (4.4), (4.3), and (4.2). In the right-hand side of the result inequality we change the function $\psi_1(\sigma)$ by its expression from (4.6). The sum obtained as a result we group by the degrees of $q/\sqrt{\sigma}$. The statement of Lemma 3.1 then follows.

Proof of Lemma 3.2.. Taking into account (2.15) and (3.1), we obtain the estimate for $K_\sigma^{(2)}(x,y)$:

$$|K_\sigma^{(2)}(x,y)| \le \int_D |K_\sigma(z,y)|\,|K_\sigma(x,z)|\,dz$$
$$\le \Big(\int_D [K_\sigma(z,y)]^2\,dz\Big)^{1/2}\Big(\int_D [K_\sigma(x,z)]^2\,dz\Big)^{1/2}$$
$$\le \gamma^2\Big(\int_D e^{-2\sigma[(z_1-x_1)^2+(z_2-x_2)^2]}\,dz\Big)^{1/2}\Big(\int_D e^{-2\sigma[(y_1-z_1)^2+(y_2-z_2)^2]}\,dz\Big)^{1/2},$$

$dz = dz_1\,dz_2\,dz_3$. Since

$$\int_D e^{-2\sigma[(z_1-x_1)^2+(z_2-x_2)^2]}\,dz \le (|D|b)^{1/2}\Big(\int_T e^{-4\sigma\alpha^2}\,dy_1\,dy_2\Big)^{1/2}$$
$$\le (|D|b)^{1/2}\Big(4\int_0^\infty\int_0^\infty e^{-4\sigma\alpha^2}\,dy_1\,dy_2\Big)^{1/2} \le \frac{\sqrt{\pi}}{2}\,\frac{(b|D|)^{1/2}}{\sqrt{\sigma}} = \frac{1}{2}\sqrt{\frac{\pi b|D|}{\sigma}},$$

then

$$|K_\sigma^{(2)}(x,y)| \le (\gamma/\sqrt{2})\sqrt{\pi b\gamma^2|D|/(2\sigma)}.$$

Continuing the process, from (2.15), for natural n and each $(x,y) \in \overline{D}$, we shall have

$$|K_\sigma^{(n+1)}(x,y)| \le (\gamma/\sqrt{2})(\sqrt{\pi b\gamma^2|D|/(2\sigma)}\,)^n.$$

As it follows from (4.7), the majorant series is converging and the first part of Lemma 3.2 follows.

Summing the majorant series, we obtain (3.8) whence (3.9) follows. Thus, Lemma 3.2 is proved. □

Proof of Lemma 3.3. The function $\Phi_\sigma(x,y)$ we represent as follows:

$$2\pi^2 e^{-\sigma(y_3^2-x_3^2)+\sigma\alpha^2}\Phi_\sigma(x,y) = I_1 + I_2,$$

where

$$I_1 = \int_0^\infty \varphi_\sigma \frac{e^{-\sigma u^2}\,du}{u^2+r^2}, \qquad I_2 = -2\int_0^\infty \varphi_\sigma \sin^2\frac{\lambda u}{2}\frac{e^{-\sigma u^2}\,du}{u^2+r^2}.$$

Taking into account (2.3), we obtain for I_1

$$I_1 = \int_0^\infty \frac{\cos\tau\sqrt{u^2+\alpha^2}}{u^2+r^2}e^{-\sigma u^2}\,du + I_1',$$

$$I_1' = -(y_3-x_3)\int_0^\infty \frac{\sin\tau\sqrt{u^2+\alpha^2}}{\sqrt{u^2+\alpha^2}}\frac{e^{-\sigma u^2}}{u^2+r^2}\,du.$$

For I_1', similarly Lemma 4.3, we obtain the estimate

$$|I_1'| \le (\pi/2 + 2\psi_0 + e^{-\sigma}/(2\sigma))/r.$$

The modulus of the first integral is less or equal to $\pi/(2r)$; therefore,

$$|I_1| \le (\pi + 2\psi_0 + e^{-\sigma}/(2\sigma))/r.$$

Analogously to Lemma 4.4, we obtain

$$|I_2| \le \frac{2}{8}\int_0^\infty u^2(2a_1^2 + a_2^2 + a_2^2\operatorname{ch} a_2 u)\Big(1 + \frac{r}{\sqrt{u^2+\alpha^2}}\Big)e^{-\sigma u^2}\frac{du}{u^2+r^2}$$
$$\le \frac{3}{8}\Big[(2a_1^2+a_2^2)\frac{\sqrt{\pi}}{2\sqrt{\sigma}} + a_2^2\frac{\sqrt{\pi}}{2\sqrt{\sigma}}e^{a_2^2/(4\sigma)}\Big].$$

Summing the estimate for $|I_1|$ and $|I_2|$ we obtain (3.10).

For $y_3 = 0$ the calculation is simplified and (3.11) is deduced easily. Calculate now $d\Phi_\sigma/dn$ and represent it in the form

$$2\pi^2 e^{\sigma\alpha^2-\sigma(y_1^2-x_1^2)} \frac{d\Phi_\sigma}{dn} = I_1 + I_2 + I_3 + I_4,$$

where

$$I_1 = -2\pi^2 e^{-\sigma\alpha^2-\sigma(y_3^2-x_3^2)}\Big(2\sigma\alpha\frac{d\alpha}{dn} - \frac{dy_3}{dn}\tau\Big)\Phi_\sigma,$$

$$I_2 = -\int_0^\infty \frac{2re^{-\sigma u^2}\cos\Theta\varphi_\sigma\cos\lambda u}{(u^2+r^2)^2}\,du, \quad I_3 = \int_0^\infty \frac{d\varphi_\sigma}{dn}\frac{e^{-\sigma u^2}}{u^2+r^2}\cos\lambda u\,du,$$

$$I_4 = \int_0^\infty \frac{\varphi_\sigma e^{-\sigma u^2}}{u^2+r^2}u\sin\lambda u\frac{d\lambda}{dn}\,du, \quad \tau = 2\sigma y_3.$$

For I_1 we obtain the estimate from (3.10):

$$|I_1| \le \Big[\pi + 2\psi_0 + \frac{e^{-\sigma}}{2\sigma}\Big]\frac{2\sigma\alpha+\tau}{r} + \frac{3}{2}(2\sigma\alpha+\tau)A_0(r); \tag{4.22}$$

the integral I_2 we represent in the form $I_2 = I_5 + I_6$, where

$$I_5 = -\int_0^\infty \frac{2r\cos\Theta\varphi_\sigma e^{-\sigma u^2}}{(u^2+r^2)^2}\,du,$$

$$I_6 = \int_0^\infty 2r\cos\Theta\varphi_\sigma\Big(2\sin^2\frac{\lambda u}{2}\Big)\frac{e^{-\sigma u^2}\,du}{(u^2+r^2)^2}.$$

Analogous to the proof of Lemmas 4.3 and 4.4, for I_5 and I_6, we obtain the estimates

$$|I_5| \le \frac{\pi}{2}\frac{|\cos\Theta|}{r^2} + \frac{2|\cos\Theta|}{r^2}\Big(\frac{\pi}{2} + 2\psi_0 + \frac{e^{-\sigma}}{2\sigma}\Big),$$

$$\begin{aligned}|I_6| &\le \frac{1}{4}\int_0^\infty u^2|\psi_\sigma|(2a_1^2 + a_2^2 + a_2^2\operatorname{ch} a_2u)e^{-\sigma u^2}\frac{r\,du}{(u^2+r^2)^2}\\
&\le \frac{1}{4}\Big[(2a_1^2+a_2^2)\int_0^\infty u^2\Big(1+\frac{r}{\sqrt{u^2+\alpha^2}}\Big)\frac{e^{-\sigma u^2}t\,du}{(u^2+r^2)^2}\\
&\quad + a_2^2\int_0^\infty u^2\Big(1+\frac{r}{\sqrt{u^2+\alpha^2}}\Big)\operatorname{ch} a_2ue^{-\sigma u^2}\frac{r\,du}{(u^2+r^2)^2}\Big]\\
&\le \frac{1}{4}\Big[(2a_1^2+a_2^2)\Big(\frac{\pi}{4}+\frac{1}{2}\Big) + a_2^2\Big(\frac{\pi}{4}+\frac{1}{2}\Big)e^{a_2^2/(4\sigma)}\Big].\end{aligned}$$

Therefore,

$$|I_2| \le \frac{\pi+2}{16}(2a_1^2 + a_2^2 + a_2^2 e^{a_2^2/(4\sigma)}) + \frac{\pi}{2}\frac{|\cos\Theta|}{r^2} + \frac{2|\cos\Theta|}{r^2}\Big(\frac{\pi}{2} + 2\psi_0 + \frac{e^{-\sigma}}{2\sigma}\Big). \quad (4.23)$$

The following inequality we obtain similarly the proof of Lemma 4.1:

$$|I_4| \le (3/4)(a_3 + a_4 + a_5)(a_1 + a_2 e^{a_2^2/(4\sigma)})\sqrt{\pi}/\sqrt{\sigma}. \quad (4.24)$$

We prove now that for each $\sigma > 0$, $x, y \in \overline{D}$ and $y \neq x$ the following inequality holds

$$|I_3| \le \frac{3}{8}(3a_2^2 + a_1^2) + \frac{3a_2^3}{16}\frac{\sqrt{\pi}}{\sqrt{\sigma}}e^{a_2^2/(4\sigma)} + \Big(3\pi + 8\psi_0 + \frac{2e^{-\sigma}}{\sigma}\Big)\sigma + [11(a_1^2 + 2a_2^2) + 3a_2^2 e^{a_2^2/(4\sigma)}]\frac{\tau\sqrt{\pi}}{32\sqrt{\sigma}} + 2\frac{\pi\tau}{r}. \quad (4.25)$$

Represent the integral I_3 in the form

$$I_3 = I_7 + I_8, \qquad I_7 = \int_0^\infty \frac{e^{-\sigma u^2}}{u^2 + r^2}\frac{d\varphi_\sigma}{dn}\,du,$$

$$I_8 = -\int_0^\infty \frac{2\sin^2(\lambda u/2)e^{-\sigma u^2}}{u^2 + r^2}\frac{d\varphi_\sigma}{dn}\,du.$$

Estimate I_8. From (2.3) we obtain

$$\frac{d\varphi_\sigma}{dn} = -\sin\tau\sqrt{u^2+\alpha^2}\,\frac{d}{dn}(\tau\sqrt{u^2+\alpha^2}\,) - \frac{dy_3}{dn}\frac{\sin\tau\sqrt{u^2+\alpha^2}}{\sqrt{u^2+\alpha^2}} - (y_3 - x_3)\Big[\frac{\cos\tau\sqrt{u^2+\alpha^2}}{\sqrt{u^2+\alpha^2}}\frac{d(\tau\sqrt{u^2+\alpha^2})}{dn} - \frac{\sin\tau\sqrt{u^2+\alpha^2}}{\sqrt{u^2+\alpha^2}}\frac{\alpha}{\sqrt{u^2+\alpha^2}}\frac{d\alpha}{dn}\Big] \quad (4.26)$$

$$\frac{d(\tau\sqrt{u^2+\alpha^2})}{dn} = 2\sigma\frac{dy_3}{dn}\sqrt{u^2+\alpha^2} + \tau\frac{\alpha}{\sqrt{u^2+\alpha^2}}\frac{d\alpha}{dn}.$$

Hence we deduce the inequality

$$\Big|\frac{d\varphi_\sigma}{dn}\Big| \le 2\sigma\sqrt{u^2+\alpha^2} + 2\tau + 2\sigma r + \frac{\tau r}{\sqrt{u^2+\alpha^2}} + \frac{\tau\alpha r}{u^2+\alpha^2}. \quad (4.27)$$

Analogously the proofs of Lemmas 4.2 and 4.3, from (4.27) we shall have

$$|I_8| \le \frac{2a_1^2+a_2^2}{4}\int_0^\infty \frac{u^2}{u^2+r^2}\Big(2\sigma\sqrt{u^2+\alpha^2}+2\tau+2\sigma r + \frac{\tau r}{\sqrt{u^2+\alpha^2}}+\frac{\tau\alpha r}{u^2+\alpha^2}\Big)e^{-\sigma u^2}\,du$$
$$+\frac{a_2^2}{4}\int_0^\infty \frac{u^2}{u^2+r^2}\Big(2\tau\sqrt{u^2+\alpha^2}+2\tau+2\sigma r + \frac{\tau r}{\sqrt{u^2+\alpha^2}}+\frac{\alpha r\tau}{u^2+\alpha^2}\Big)\operatorname{ch} a_2 u\, e^{-\sigma u^2}\,du,$$

or

$$|I_8| \le \frac{2a_2^2+a_1^2}{8}\Big(3+\frac{11}{4}\frac{\tau\sqrt{\pi}}{\sqrt{\sigma}}\Big)+\frac{a_2^2}{4}\Big(\frac{3}{2}+\frac{3a_2}{4}\frac{\sqrt{\pi}}{\sqrt{\sigma}}e^{a_2^2/(4\sigma)}+\frac{3\tau}{8}\frac{\sqrt{\pi}}{\sqrt{\sigma}}e^{a_2^2/(4\sigma)}\Big). \tag{4.28}$$

From (4.26) we see that I_7 is represented as the sum of six integrals. The first term we denote by I_9:

$$I_9 = 2\sigma\frac{dy_3}{dn}\int_0^\infty \frac{\sin\tau\sqrt{u^2+\alpha^2}}{\sqrt{u^2+\alpha^2}}\,\frac{(u^2+\alpha^2)e^{-\sigma u^2}}{u^2+r^2}\,du.$$

In the right-hand side the Dirichlet integral stands which we had estimated before. Therefore,

$$|I_9| \le 2\sigma(\pi+4\psi_0+e^{-\sigma}/\sigma).$$

The rest integrals are estimated easily, and for I_7 we deduce the inequality:

$$|I_7| \le 2\pi\tau/r+\pi\sigma+2\sigma(\pi+4\psi_0+e^{-\sigma}/\sigma). \tag{4.29}$$

Estimates (4.29) and (4.28) yield (4.25). Now, from (4.25) we obtain (3.13).
Further, we calculate $d\Phi_\sigma/dy_3$ and set $y_3=0$. In this case the calculations are simplified and we obtain (3.16). Lemma 3.3 is proved. □

REFERENCES

1. L. A. Aizenberg, *The Carleman Formulas in the Complex Plane.* Nauka, Novosibirsk, 1990 (in Russian).
2. K. Miranda, *Equations with Partial Derivatives of Elliptic Type.* IL, Moscow, 1957 (in Russian).
3. M. M. Lavrent'ev, *On Certain Ill-Posed Problems of Mathematical Physics.* Nauka, Novosibirsk, 1962 (in Russian).

4. M. M. Lavrent'ev, On the Cauchy problem for Laplace equation. *Izv. Akad. Nauk SSSR. Ser. Mat.* (1956) **20**, 819–842 (in Russian).

5. I. Islomov, On construction of the regularized solution of the Cauchy problem for the Helmholtz equation. *Dokl. Uzb. SSR* (1984) **8**, 10–11.

6. I. Islomov, On the Cauchy problem for the Helmholtz equation. In: *Uniqueness, Stability and Methods of Solutions of Ill-Posed Problems of Mathematical Physics and Analysis.* Novosibirsk, 1984, 97–105 (in Russian).

7. Sh. Yarmukhamedov, On the Cauchy problem for the Laplace equation. *Dokl. Akad. Nauk SSSR* (1977) **235**, No. 2, 281–283 (in Russian).

8. Sh. Yarmukhamedov, On the harmonic extension of continuous functions defined on the piece of the boundary. *Acad. Sci. Dokl. Math.* (1993) **46**, No. 3, 430–433 (in Russian).

9. Sh. Yarmukhamedov, On extension of solution of the Helmholtz equation. *Dokl. Ros. Akad. Nauk* (1997) **357**, No. 5, 320–323 (in Russian).

10. Sh. Yarmukhamedov, On harmonic extension of differentiable functions set in the piece of the boundary. *Sib. Mat. J.* (2002) **43**, No. 1, 228–239 (in Russian).

Ill-Posed and Non-Classical Problems of
Mathematical Physics and Analysis, Samarkand, 2000, pp. 173–178
M. M. Lavrent'ev and S. I. Kabanikhin (Eds)

On problem for a third order equation with multiple characteristics

S. ABDINAZAROV and B. M. KHOLBOEV

Abstract — The goal of this paper is investigation of the equation

$$Lu = \frac{\partial^3 u}{\partial x^3} - \frac{\partial u}{\partial y} = f(x, y) \tag{1}$$

in the domain $D = \{(x, y) \mid h_1(y) < x < h_2(y),\ 0 < y \leq Y\}$ with the boundary conditions

$$u(x, 0) = F(x), \qquad h_1(0) \leq x \leq h_2(0), \tag{2}$$

$$u_x(h_1(y), y) = \varphi_1(y), \qquad u_{xx}(h_1(y), y) = \varphi_2(y),$$
$$u_x(h_2(y), y) = \varphi_3(y), \qquad 0 \leq y \leq 1, \tag{3}$$

where $f(x, y)$, $\varphi_i(y) (i = \overline{1, 3})$, $F(x)$ are given functions. These functions satisfy the following concordance conditions

$$F'(h_1(0)) = \varphi_1(0), \quad F''(h_1(0)) = \varphi_2(0), \quad F'(h_2(0)) = \varphi_3(0). \tag{4}$$

The analogous research for equation (1) with the other boundary conditions was done in [1–3].

1. UNIQUENESS OF SOLUTION

Theorem 1. *If $h_i(y) \in C'([0, 1])$, $i = 1, 2$ then the solution of problem* (1)–(3) *is unique.*

Proof. Suppose there exist two solutions of the problem in question $u_1(x, y)$ and $u_2(x, y)$.

We set $v(x,y) = u_1(x,y) - u_2(x,y)$. Then, for the function $v(x,y)$ we obtain the following problem

$$\begin{gathered} L(v) = 0, \\ v(x,0) = 0, \qquad h_1(0) \le x \le h_2(0), \\ v_x(h_1(y),y) = 0, \qquad v_{xx}(h_1(y),y) = 0, \\ v_x(h_2(y),y) = 0, \qquad 0 \le y \le 1, \end{gathered} \tag{5}$$

Consider the identity

$$\int\int_D v_{xx}(v_{xxx} - v_y)\,\mathrm{d}x\,\mathrm{d}y = 0. \tag{6}$$

Integrating by parts and using the homogeneous conditions (5) we obtain

$$\frac{1}{2}\int_0^1 v_{xx}^2(h_2(y),y)\,\mathrm{d}y + \frac{1}{2}\int_{h_1(1)}^{h_2(1)} v_x^2(x,1)\,\mathrm{d}x = 0. \tag{7}$$

Hence, we have

$$v_{xx}(h_2(y),y) = 0, \qquad v_x(x,1) = 0.$$

Therefore, $v(x,1) = \text{const}$.

Suppose now that

$$v(x,y) = \omega(x,y)e^{My}, \qquad M = \text{const} \ne 0.$$

Then we obtain

$$\iint_D \omega_{xx} L\omega\, e^{My}\,\mathrm{d}x\,\mathrm{d}y = \frac{M}{2}\iint_D \omega_x^2\, e^{My}\,\mathrm{d}x\,\mathrm{d}y = 0,$$

where

$$L\omega = \omega_{xxx} - \omega_y - M\omega. \tag{8}$$

Therefore, $w_x(x,y) = 0$ and $\omega(x,y) = w(y)$. Substituting the function $w(y)$ into (2) and (8), we have

$$w'(y) - Mw(y) = 0, \qquad w(0) = 0. \tag{9}$$

It is known that problem (9) has the trivial solution only. Therefore, $v(x,y) = 0$ in $\overline{D}$. $\square$

2. EXISTENCE OF SOLUTION

Let $F(x) \in C^3(h_1(0), h_2(0))$. Without loss of generality we may set $F(x) = 0$.

Really, if we set

$$u(x,y) = v(x,y) + F(x), \tag{10}$$

then, for the function $v(x,y)$, we shall obtain the following problem

$$\begin{gathered} L(v) = \bar{f}(x,y), \qquad v(x,0) = 0, \qquad h_1(0) \le x \le h_2(0), \\ v_x(h_1(y),y) = \overline{\varphi}_1(y), \qquad v_{xx}(h_1(y),y) = \overline{\varphi}_2(y), \\ v_x(h_2(y),y) = \overline{\varphi}_3(y), \qquad 0 \le y \le 1, \end{gathered} \tag{11}$$

where

$$\begin{aligned} \overline{\varphi}_1(y) &= \varphi_1(y) - F'(h_1(y)), \qquad & \overline{\varphi}_2(y) &= \varphi_2(y) - F''(h_1(y)), \\ \overline{\varphi}_3(y) &= \varphi_3(y) - F'(h_2(y)), \qquad & \bar{f}(x,y) &= f(x,y) - F'''(x). \end{aligned}$$

As is shown in [1], the function

$$W(x,y) = \frac{1}{\pi} \iint_D U(x,y;\xi,\eta) \bar{f}(\xi,\eta)\, \mathrm{d}\xi\, \mathrm{d}\eta \tag{12}$$

satisfies equation (1) and the condition $w(x,0) = 0$ if $\bar{f}(\xi,\eta) \in C^{0,1}_{x,y}(\overline{D})$, $f(x,0)$. Here $U(x,y;\xi,\eta)$ is the fundamental solution of equation (1) (see [1]) which is as follows:

$$U(x,y;\xi,\eta) = \begin{cases} \dfrac{1}{(y-\eta)^{1/3}} f\Big(\dfrac{x-\xi}{(y-\eta)^{1/3}}\Big), & y > \eta,\ x \ne \xi, \\ 0, & y \le \eta, \end{cases}$$

where

$$f(t) = \int_0^\infty \cos(\lambda^3 - \lambda t)\, \mathrm{d}\lambda, \qquad t = \frac{x-\xi}{(y-\eta)^{1/3}}.$$

The function $f(t)$ is called the Airy function and satisfies the equation

$$z''(t) + tz(t)/3 = 0. \tag{13}$$

The function $f(t)$ satisfies the following relations (see [4]):

$$\bar{f}^{(n)}(t) \sim C_n^+ t^{n/2-1/4} \sin(2t^{3/2}/3) \qquad \text{for} \quad t \to +\infty, \tag{14}$$

$$\bar{f}^{(n)}(t) \sim C_n^- t^{n/2-1/4} \sin(-2t^{3/2}/3) \qquad \text{for} \quad t \to -\infty, \tag{15}$$

where C_n^- and C_n^+ are constants.

Taking into account these relations, we shall seek for the solution of problem (11) in the form

$$v(x,y) = \omega(x,y) + w(x,y). \tag{16}$$

Then, relatively the function $\omega(x,y)$, we obtain the following problem

$$L\omega = 0, \tag{17}$$

$$\begin{gathered} \omega(x,0) = 0, \qquad h_1(0) \le x \le h_2(0), \\ \omega_x(h_1(y),y) = \tilde{\varphi}(y), \qquad \omega_{xx}(h_1(y),y) = \tilde{\varphi}_2(y), \\ \omega_x(h_2(y),y) = \tilde{\varphi}_3(y), \qquad 0 \le y \le 1, \end{gathered} \tag{18}$$

where

$$\begin{gathered} \tilde{\varphi}_1(y) = \tilde{\varphi}_1(y) - W_x(h_1(y),y), \\ \tilde{\varphi}_2(y) = \tilde{\varphi}_2(y) - W_{xx}(h_1(y),y), \qquad \tilde{\varphi}_3(y) = \tilde{\varphi}_3(y) - W_x(h_2(y),y). \end{gathered}$$

Solution of problem (17), (18) we seek in the form

$$\omega(x,y) = \int_0^y U(x,y;0,\eta)\alpha_1(\eta)\,d\eta + \int_0^y U(x,y;1,\eta)\alpha_2(\eta)\,d\eta + \int_0^y V(x,y;0,\eta)\alpha_3(\eta)\,d\eta. \tag{19}$$

Here

$$V(x,y;\xi,\eta) = \begin{cases} \dfrac{1}{(y-\eta)^{1/3}}\,\varphi\Big(\dfrac{x-\xi}{(y-\eta)^{1/3}}\Big), & y > \eta,\ x > \xi, \\ 0, & y \le \eta \end{cases}$$

$$\varphi(t) = \int_0^\infty [\exp(-\lambda^3 - \lambda t) + \sin(\lambda^3 - \lambda t)]\,d\lambda, \qquad t = \frac{x-\xi}{(y-\eta)^{1/3}}.$$

Note that $\varphi(t)$ satisfies the relations (13), (14), and (15) also. Satisfying the boundary conditions (18) we obtain

$$\tilde{\varphi}_1(y) = \int_0^y U_x(h_1(y),y;h_1(\eta),\eta)\alpha_1(\eta)\,d\eta + \int_0^y U_x(h_1(y),y;h_2(\eta),\eta)\alpha_2(\eta)\,d\eta + \int_0^y V_x(h_1(y),y;h_1(\eta),\eta)\alpha_3(\eta)\,d\eta, \tag{20}$$

$$\tilde{\varphi}_2(y) = \int_0^y U_{xx}(h_1(y), y; h_1(\eta), \eta)\alpha_1(\eta)\,\mathrm{d}\eta + \int_0^y U_{xx}(h_1(y), y; h_2(\eta), \eta)\alpha_2(\eta)\,\mathrm{d}\eta + \int_0^y V_{xx}(h_1(y), y; h_1(\eta), \eta)\alpha_3(\eta)\,\mathrm{d}\eta, \quad (21)$$

$$\tilde{\varphi}_3(y) = \int_0^y U_x(h_2(y), y; h_1(\eta), \eta)\alpha_1(\eta)\,\mathrm{d}\eta + \int_0^y U_x(h_2(y), y; h_2(\eta), \eta)\alpha_2(\eta)\,\mathrm{d}\eta + \int_0^y V_x(h_2(y), y; h_1(\eta), \eta)\alpha_3(\eta)\,\mathrm{d}\eta. \quad (22)$$

Applying the Abel transform and after non-tedious computation we obtain the following system

$$\int_0^z \frac{\tilde{\varphi}_1''(y)}{(z-y)^{1/3}}\,\mathrm{d}y = \frac{2\pi}{\sqrt{3}}\, f'(0)\alpha_1'(z) + \frac{2\pi}{\sqrt{3}}\,\varphi'(0)\alpha_3'(z) + \int_0^z \alpha_1'(\eta)\,\mathrm{d}\eta \frac{\partial}{\partial z}\int_\eta^z \Big[\frac{\partial}{\partial z}\int_t^z \frac{1}{(z-y)^{1/3}}\,\frac{\partial U(h_1(y), y; h_1(t), t)}{\partial x}\,\mathrm{d}y\Big]\,\mathrm{d}t + \int_0^z \alpha_3'(\eta)\,\mathrm{d}\eta\,\frac{\partial}{\partial z}\int_\eta^z \Big[\frac{\partial}{\partial z}\int_t^z \frac{1}{(z-y)^{1/3}}\,\frac{\partial V(h_1(y), y; h_1(t), t)}{\partial x}\,\mathrm{d}y\Big]\,\mathrm{d}t + \int_0^z \alpha_2(\eta)\,\mathrm{d}\eta \frac{\partial}{\partial z}\int_\eta^z \frac{1}{(z-y)^{1/3}}\,\frac{\partial^2 U(h_1(y), y; h_2(t), t)}{\partial x \partial y}\,\mathrm{d}y, \quad (23)$$

$$\tilde{\varphi}_2'(z) = \frac{\pi}{3}\,\alpha_1'(z) + \int_0^z \alpha_1'(\eta)\,\mathrm{d}\eta\,\frac{\partial}{\partial z}\int_\eta^z \frac{\partial^2 U(h_1(z), z; h_1(t), t)}{\partial x^2}\,\mathrm{d}t + \int_0^z \alpha_2(\eta)\,\frac{\partial}{\partial z}\Big(\frac{\partial^2 U(h_1(z), z; h_2(\eta), \eta)}{\partial x^2}\Big)\,\mathrm{d}\eta, \quad (24)$$

$$\int_0^z \frac{\tilde{\varphi}_3''(y)}{(z-y)^{1/3}}\,\mathrm{d}y = \int_0^z \alpha_1'(\eta)\,\mathrm{d}\eta \times \frac{\partial}{\partial z}\int_\eta^z \Big[\frac{\partial}{\partial z}\int_t^z \frac{1}{(z-y)^{1/3}}\,\frac{\partial U(h_2(y), y; h_1(t), t)}{\partial x}\,\mathrm{d}y\Big]\,\mathrm{d}t + \int_0^z \alpha_2(\eta)\,\mathrm{d}\eta\,\frac{\partial}{\partial z}\int_\eta^z \frac{1}{(z-y)^{1/3}}\,\frac{\partial^2 U(h_2(y), y; h_2(t), t)}{\partial x \partial y}\,\mathrm{d}y + \int_0^z \alpha_3'(\eta)\,\mathrm{d}\eta \frac{\partial}{\partial z}\int_\eta^z \Big[\frac{\partial}{\partial z}\int_t^z \frac{1}{(z-y)^{1/3}}\,\frac{\partial V(h_2(y), y; h_1(t), t)}{\partial x}\,\mathrm{d}y\Big]\,\mathrm{d}t. \quad (25)$$

The system (23)–(25) is equivalent to the system of integral Volterra equations of the second kind

$$\alpha_l(z) = R_l(z) + \sum_{s=1}^{3} \int_0^z N_{sl}(z,\eta)\alpha_3(\eta)\, d\eta, \qquad l = \overline{1,3}, \tag{26}$$

where $\alpha_l(z)$ are the desired functions, $R_l(z)$ are known and are expressed via the functions $\tilde{\varphi}_i(z)$ $(i = \overline{1,3})$. $N_{sl}(z,\eta)$ is the matrix whose elements are expressed via the fundamental solution of equation (18).

It is easy to show that the kernels $N_{sl}(z,\eta)$ have the weak singularity:

$$|N_{sl}(z,\eta)| < C/|z-\eta|^{1/2}; \qquad c = \text{const} > 0. \tag{27}$$

From the general theory of Volterra equations [5] it follows that system (26) is solvable uniquely in the class of continuous functions which admit the representation of the form

$$al_l(z) = R_l(z) + \sum_{j=0}^{3} \int_0^z H_{sl}(z,\eta)R_s(\eta)\, d\eta; \quad l = \overline{1,3}.$$

Here the resolvent $H_{sl}(z,\eta)$ has a weak singularity of type (27).

REFERENCES

1. L. Cattabriga, Annali della seuola narmole. *Supericri di Pisa e Mat.* (1959) **13**, No. 2, 163.

2. T.D. Dzhuraev, *The Boundary-Value Problems for the Equations of Mixed Type and of Mixed-Composite Type.* FAN, Tashkent, 1979 (in Russian).

3. S. Abdinazarov, General boundary-value problems for the third order equation with multiply characteristics. *Differential Equations* (1981) **17**, No. 1, Moscow, 3–12 (in Russian).

4. V.M. Fedoryuk, *Saddle-Point Method.* Nauka, Moscow, 1977 (in Russian).

5. M. Gervey. *J. Math. Pures et Appl.* (1913) **9**, No. 1–4, 305.

Ill-Posed and Non-Classical Problems of Mathematical Physics and Analysis, Samarkand, 2000, pp. 179–183
M. M. Lavrent'ev and S. I. Kabanikhin (Eds)

On uniqueness and stability of solution of the Cauchy problem for pseudoparabolic equation

B. K. AMONOV* and S. S. KOBILOV*

Stability and uniqueness play an important role when researching well-posedness of problems of mathematical physics.

Lavrent'ev, M. M. [1] had established stability of determination of a harmonic function in the plane or spatial domain given the Cauchy data in the piece of the boundary in the class of bounded functions. The stability estimates for solutions of general elliptic equations are obtained in [2]. The contents of these papers and some other results can be found in the monograph [3]. The uniqueness theorems for the noncharacteristic Cauchy problem for the general parabolic equation of the second order is established first in [6].

In our paper the uniqueness theorem is established and the stability estimate is obtained for the Cauchy problem for the pseudoparabolic equation.

Suppose that the function $u(x,t)$ is continuous in the domain

$$\Pi = \{(x,t) \mid 0 \le x \le \pi, \ \ 0 \le t < \infty\}.$$

Inside this domain the function $u(x,t)$ satisfies the equation

$$u_t(x,t) = u_{xxt}(x,t) + u_{xx}(x,t) \tag{1}$$

and the conditions

$$u(0,t) = 0, \qquad u(\pi,t) = 0 \qquad (t \ge 0) \tag{2}$$

$$u'_x(0,t) = f(t) \qquad (t \ge 0). \tag{3}$$

*Samarkand State University, University prosp., 15, Samarkand 703004, Uzbekistan

Here $f(t)$ is a continuous function and

$$\int_0^{\pi} u^2(x,0)\,\mathrm{d}x \le M^2, \qquad M > 0. \tag{4}$$

We need to recover the function $u(x,t)$, given $f(t)$; to establish the uniqueness theorem and to obtain the stability estimate.

Theorem (Uniqueness theorem). *If $f(t) = 0$ then $u(x,t) \equiv 0$ in the domain Π.*

Proof. It is easy to verify that the solution of equation (1) satisfying the conditions (2)–(3) has the form

$$u(x,t) = \sum_{k=1}^{\infty} a_k e^{k^2 t/(1+k^2)} \sin kx, \tag{5}$$

where a_k are the Fourier coefficients. Relations (3) and (5) yield

$$\sum_{k=1}^{\infty} a_k e^{k^2 t/(1+k^2)} = f(t). \tag{6}$$

Let $F(p)$ be the Laplace transform of the function $f(t)$, i.e.

$$F(p) = \int_0^{\infty} f(t) e^{-pt}\,\mathrm{d}t. \tag{7}$$

Applying the Laplace transformation with respect to t to both sides of equality (6) we find

$$\sum_{k=1}^{\infty} k a_k \frac{1}{k^2/(1+k^2)+p} = F(p). \tag{8}$$

Integrate both sides of this equality along the circles

$$\Gamma_n = \Big\{ p \mid |p - p_n| = \frac{1}{2}\Big[\frac{(n+1)^2}{(n+1)^2+1} - \frac{n^2}{n^2+1}\Big]\Big\}$$

with the centers in the points $p_n = -n^2/(n^2+1)$. As a result we obtain

$$\sum_{k=1}^{\infty} k a_k \int_{\Gamma_n} \frac{\mathrm{d}p}{k^2/(1+k^2)+p} = \int_{\Gamma_n} F(p)\,\mathrm{d}p.$$

Taking into account the Cauchy formula

$$f(z) = \frac{1}{2\pi i}\int_C \frac{f(\xi)\,\mathrm{d}\xi}{\xi - z}$$

we have

$$na_n 2\pi i = \int_{\Gamma_n} F(p)\,\mathrm{d}p.$$

Hence we find the coefficients

$$a_n = \frac{1}{2n\pi i}\int_{\Gamma_n} F(p)\,\mathrm{d}p.$$

Thus, the Fourier coefficients are defined uniquely. Therefore, the result of the uniqueness theorem follows.

Consider now the stability of solution.

Suppose $|f(t)| < \varepsilon$, $0 \le t < \infty$. Formula (5) yields the expression for the function $f(t)$:

$$f(t) = u'_x(0,t) = \sum_{k=1}^{\infty} k a_k e^{-k^2 t/(1+k^2)}. \tag{9}$$

Setting in this equality $e^{-t} = z$ we obtain

$$\Phi(z) = \sum_{k=1}^{\infty} k a_k z^{k/(1+k^2)}. \tag{10}$$

Since $e^{-t} = z$ and $0 \le t < \infty$, the function $\Phi(z)$ is defined in the segment $0 \le z \le 1$,

$$|\Phi(z)| < \varepsilon \qquad \text{for} \qquad 0 \le z \le 1 \tag{11}$$

and inequality (4) holds. We try now to estimate the solution of problem (1)–(3). Condition (4) yields

$$\sum_{k=1}^{\infty} a_k^2 \le M^2. \tag{12}$$

Denote by D_ρ the disk $|z| \le \rho < 1$. Relations (10) and (12) yield the inequality

$$\begin{aligned} |\Phi(z)|_{|z|\le\rho<1} &\le \sum_1^{\infty} K|a_k|\rho^{k^2} \le \sum_1^{\infty} k|a_k|\rho^k \\ &\le \Big(\sum_1^{\infty}|a_k|^2\Big)^{1/2}\Big(\sum_1^{\infty} k^2\rho^{2k}\Big)^{1/2} \le C(\rho)M, \end{aligned}$$

where $C(\rho)$ is a constant dependent on ρ only. Now, we map the disk $\{z \mid |z| \le 1\}$ onto the half-disk $\{w \mid |w| < 1, \operatorname{Im} w \ge 0\}$ conformally using the function $w = \sqrt{z}$. In this case D_ρ passes to the half-disk

$$\mathrm{d}_{\rho^{1/2}} = \{w \mid w \le \rho^{1/2}, \ \operatorname{Im} w \ge 0\}.$$

Denote by $w_\rho(z)$ the harmonic measure of diameter relatively the half-disk $\mathrm{d}_{\rho^{1/2}}$. It is known [7] that

$$w_\rho(z) = 1 - \frac{2}{\pi} I_m \ln \frac{\rho^{1/2} - z}{\rho^{1/2} + z}.$$

Since $|\Phi(z)| < \varepsilon$ in $[-\rho^{1/2}, \rho^{1/2}]$ and $|\Phi(z)| \le C(\rho)M$in the half-circle $\{w \mid |w| = \rho^{1/2}, \operatorname{Im} w > 0\}$, then, taking into account the theorem on the two constants (R. Nevanlinna), we have

$$|\Phi(z)| \le \varepsilon^{w_\rho(z)} [MC(\rho)]^{1-w_\rho(z)},$$

or

$$\sum_1^\infty k^2 |a_k|^2 |z|^{2k^2} \le \varepsilon^{2w_\rho(z)} [MC(\rho)]^{1-w_\rho(z)}.$$

Hence, on the bases of the Hölder inequality, we obtain

$$\begin{aligned} |u(x,z)| &\le \sum_1^\infty \frac{k|a_k|}{k} |z|^{k^2} \le \Big(\sum_1^\infty k^2 |a_k|^2 |z|^{2k^2}\Big)^{1/2} \sum_1^\infty \Big(\frac{1}{k^2}\Big)^{1/2} \\ &\le \varepsilon^{w_\rho(z)} [MC(\rho)]^{1-w_\rho(z)} C_1, \end{aligned}$$

where C_1 is a constant.

We have obtained the stability estimate for solution of problem (1)–(3). □

REFERENCES

1. M. M. Lavrent'ev, On the Cauchy problem for the Laplace equation. In: *Trudy 3 Vsesousn. Math. S'ezda.* Moscow, 1956, **2** (in Russian).

2. M. M. Lavrent'ev, On the Cauchy problem for elliptic equations of the second order. *Dokl. Acad. Nauk SSSR* (1957) **112**, No. 2, 195–197 (in Russian).

3. M. M. Lavrent'ev, *On Certain Ill-Posed Problems of Mathematical Physics.* Izd. Sib. Otdel. Akad. Nauk SSSR, Novosibirsk, 1962 (in Russian).

4. S. Mizohata, Unicite du prolongement des solutions pour quelques operateurs differentiels paraboliquers. *Mem. Coll. Sci. Univ. Kyoto. Sec. A* (1958) **31**, 219–239.

5. M. H. Protter, Properties of solutions of parabolic equations and inequalities. *Canad. J. Math.* (1961) **13**, 331–345.

6. B. K. Amonov and S. P. Shishatskii, A priori estimate for solution of the Cauchy problem with the data on a time-like surface for second order parabolic equation and the uniqueness theorem connected with it. *Dokl. Akad. Nauk SSSR* (1972) **206**, No. 1, 11–12 (in Russian).

7. M. A. Evgrafov, *Analytical Functions.* Nauka, Moscow, 1968 (in Russian).

Ill-Posed and Non-Classical Problems of
Mathematical Physics and Analysis, Samarkand, 2000, pp. 185–190
M. M. Lavrent'ev and S. I. Kabanikhin (Eds)

Uniqueness theorem for an unbounded domain

Z. R. ASHUROVA* and Yu. J. ZHURAEV*

Abstract — In this paper the uniqueness theorem is established for a class of harmonic functions defined in an unboundary domain which lies in a layer.

1. INTRODUCTION

We suppose that $\mathcal{D}$ is unbounded domain from $\mathbb{R}^3$ lying inside the layer

$$0 \leq y_3 \leq h, \quad h = \pi/\rho; \quad y = (y_1, y_2, y_3).$$

Suppose that $\partial\mathcal{D} = \Gamma_1 \cup \Gamma_2$ are the smooth surfaces given by the equations $y_3 = f_i(y_1, y_2)$, $i = 1, 2$, so that

$$|f_i(y_1, y_2)| \leq \text{const}, \quad \left|\frac{\partial f_i}{\partial y_j}(y_1, y_2)\right| \leq \text{const}, \quad j = 1, 2; \ \ i = 1, 2.$$

Let $\mathcal{A}_\rho(\mathcal{D})$ be the space consisted of the functions harmonic in $\mathcal{D}$ and continuous together with their first partial derivatives up to the finite points of the boundary $\partial\mathcal{D}$ and

$$|U(y)| + |\operatorname{grad} U(y)| \leq \exp\left(o(\exp \rho |y'|)\right), \quad |y'| = \sqrt{y_1^2 + y_2^2} \to \infty.$$

In our paper we shall prove the following theorem.

Theorem. *If a function $U \in \mathcal{A}_\rho(\mathcal{D})$ satisfies the conditions*

$$U(y) = 0, \quad y \in \partial\mathcal{D}, \tag{1.1}$$

*Samarkand State University, Samarkand 703004, Uzbekistan

$$\left|\frac{\partial U}{\partial n}(y)\right| \leq \exp\left(c|y|\right), \quad y \in \partial\mathcal{D}, \tag{1.2}$$

(where n is the vector external normal to $\partial\mathcal{D}$ and c is a certain constant), then $U \equiv 0$ in $\mathcal{D}$.

In the case when condition (1.2) is replaced by the condition

$$\left|\frac{\partial U}{\partial n}(y)\right| \leq c_1|y|^k, \qquad c_1 = \text{const}, \quad y \in \partial\mathcal{D},$$

where k is a nonnegative integer the theorem was proved by Ashurova [1]. Our proof will use the integral representation of harmonic functions in unbounded domains with a noncompact boundary [3] and the theorem of Landis [2].

2. THE BASIC INEQUALITY

Let $y = (y_1, y_2, y_3)$, $x = (x_1, x_2, x_3)$, $y \in \mathbb{R}^3$, $0 < x_3$, $y_3 < h$, $h = \pi/\rho$, where $\rho > 0$. For $x \neq y$ we can define the function $\Phi(y, x)$ by the formula

$$\Phi(y,x) = -\frac{1}{2\pi^2 K(x_3)} \int_0^\infty \operatorname{Im} \frac{K(y_3 + i\eta)}{y_3 - x_3 + i\eta} \frac{du}{\eta}, \tag{2.1}$$

where $\eta^2 = u^2 + \alpha^2$, $\alpha^2 = (y_1 - x_1)^2 + (y_2 - x_2)^2$,

$$K(\omega) = (\omega + 3h - x_3)^{-1} \exp\left(-a \operatorname{ch} i\rho_1(\omega - h/2)\right), \qquad \omega = \xi + i\eta. \tag{2.2}$$

Here a, ρ_1 are positive and $0 < \rho_1 < \rho$. Then, the following estimate holds for the function $\Phi(y, x)$:

$$|\Phi(y,x)| \leq \frac{C_0}{r \exp\left(a \cos(\rho_1(y_3 h/2)) \operatorname{ch} \rho_1 \alpha\right)}. \tag{2.3}$$

Here $r^2 = \alpha^2 + (y_3 - x_3)^2$ and $C_0 = \text{const}$. In this section we shall establish inequality (2.3). If $\omega = x_3$, then, following (2.2), we obtain

$$K(x_3) = (3h)^{-1} \exp\left(-a \cos \rho_1(x_3 - h/2)\right).$$

Since $-\pi/2 < \rho_1(x_3 - h/2) < \pi/2$, then $\cos \rho_1(x_3 - h/2) > 0$; therefore,

$$1/K(x_3) = 3h \exp\left(a \cos \rho_1\, x_3 h/2\right) \leq 3h \exp a. \tag{2.4}$$

Set $\beta_1 = y_3 - x_3 + 3h$, $r_1^2 = \alpha^2 + \beta_1^2$. It can be shown that

$$\operatorname{Im} \frac{K(y_3 + i\eta)}{(y_3 - x_3 + i\eta)} = \frac{(\beta\beta_1 - \eta^2)\sin a(\sin\rho_1(y_3 - h/2)\operatorname{sh}\rho_1\eta)}{(u^2 + r^2)(u_2 + r_1^2)\exp(a\cos\rho_1(y_3 - h/2)\operatorname{ch}\rho_1\eta)} - \frac{\eta(\beta + \beta_1)\cos(a\sin\rho_1(y_3 - h/2)\operatorname{sh}\rho_1\eta)}{(u^2 + r^2)(u_2 + r_1^2)\exp(a\cos\rho_1(y_3 - h/2)\operatorname{ch}\rho_1\eta)}. \quad (2.5)$$

Since $\alpha \le \eta$, then

$$\exp(a\cos\rho_1(y_3 - h/2)\operatorname{ch}\rho_1\alpha) \le \exp(a\cos\rho_1(y_3 - h/2)\operatorname{ch}\rho_1\eta).$$

Taking into account that

$$\lim_{\eta\to 0} \frac{\beta\beta_1\sin(a\operatorname{sh}\rho_1\eta)}{\eta^2 + \beta^2} = 0,$$

we see that there exists such $\varepsilon > 0$, that for $|\eta| < \varepsilon$, the function $\beta\beta_1\sin(a\operatorname{sh}\rho_1\eta)/(\eta^2 + \beta^2)$ is bounded. If $|\eta| \ge \varepsilon$ then we have the inequality

$$\left|\frac{\beta\beta_1\sin(a\operatorname{sh}\rho_1\eta)}{\eta^2 + \beta^2}\right| \le \frac{4h}{\varepsilon^2}.$$

Besides, we have

$$\frac{\eta^2}{u^2 + r^2} = \frac{\eta^2}{\eta^2 + \beta^2} \le 1.$$

As a result we obtain

$$\left|\frac{(\beta\beta_1 - \eta^2)\sin a(\sin\rho_1(y_3 - h/2)\operatorname{sh}\rho_1\eta)}{(u^2 + r^2)\exp(a\cos\rho_1(y_3 - h/2)\operatorname{ch}\rho_1\eta)}\right| \le \frac{C_1}{\exp(a\cos\rho_1(y_3 - h/2)\operatorname{ch}\rho_1\alpha)}, \quad (2.6)$$

where C_1 is a certain constant. (Further, by C_i we shall denote the constants). The following estimate can be obtained easily:

$$\left|\frac{\eta(\beta + \beta_1)\cos(a\sin\rho_1(y_3 h/2)\operatorname{sh}\rho_1\eta)}{\exp(a\cos\rho_1(y_3 - h/2)\operatorname{ch}\rho_1\eta)}\right| \le \frac{C_2\eta}{\exp(a\cos\rho_1(y_3 h/2)\operatorname{ch}\rho_1\alpha)}. \quad (2.7)$$

At first we shall estimate the function $\Phi(y,x)$. The inequalities (2.4)–(2.7) yield the estimate

$$|\Phi(y,x)| \le \frac{C_3}{\exp(\cos\rho_1(y_3-h/2)\operatorname{ch}\rho_1\alpha)} \times \left(\int_0^\infty \frac{du}{(u^2+r^2)\sqrt{u^2+r^2}} + \int_0^\infty \frac{du}{(u^2+r^2)(u^2+r_1^2)}\right).$$

Since

$$u/\sqrt{u^2+\alpha^2} \le 1,$$

the integral

$$\int_0^\infty \frac{du}{u^2+r_1^2}$$

is converging and

$$\int_0^\infty \frac{du}{(u^2+r^2)(u^2+r_1^2)} \le \frac{C}{r},$$

we have

$$|\Phi(y,x)| \le \frac{C_0}{r\exp(a\cos\rho_1(y_3 h/2)\operatorname{ch}\rho_1\alpha)}.$$

Thus, inequality (2.3) is established.

3. PROOF OF THE THEOREM

Under the conditions of the theorem, following [3], we have

$$\int_{\partial\mathcal{D}}\left(U\frac{\partial\Phi}{\partial n} - \Phi\frac{\partial U}{\partial n}\right)ds = \begin{cases} 0, & x \notin \mathcal{D}\bigcup\partial\mathcal{D} \\ U(x), & x\in\mathcal{D}. \end{cases}$$

Taking into account (1.1), we can rewrite this formula as follows:

$$U(x) = \int_{\Gamma_1}\Phi(y,x)\frac{\partial U}{\partial n}\,ds \int_{\Gamma_2}\Phi(y,x)\frac{\partial U}{\partial n}\,ds, \quad x\in\mathcal{D}. \tag{3.1}$$

Introduce the notations

$$I_j = -\int_{\Gamma_j}\Phi(y,x)\frac{\partial U}{\partial n}\,ds, \quad j=1,2, \quad a_1 = a\cos\rho_1\frac{h}{2}.$$

Since Γ_i is given by the equation $y_3 = f_1(y_1,y_2)$, where f_1 is bounded together with its partial derivatives, inequalities (1.2) and (2.3) yield

$$|I_1| \le C_4\int\int_{\mathbb{R}^2}\frac{\exp(c\sqrt{y_1^2+y_2^2+h^2})}{r\exp(a_1\operatorname{ch}\rho_1\alpha)}\,dy_1\,dy_2 = C_4(I_{11}+I_{12}+I_{13}+I_{14}),$$

where by I_{1i}, $i = 1,2,3,4$ we denote the integrals along the quadrants respectively. First we estimate I_{11}. Represent it in the form

$$I_{11} = \int_0^\infty \int_0^\infty \frac{\exp\,(c\sqrt{y_1^2 + y_2^2 + h^2})}{r\varepsilon \exp\,(a_1 \operatorname{ch} \rho_1 \alpha)}\, \mathrm{d}y_1\, \mathrm{d}y_2. \tag{3.2}$$

As $\sqrt{y_1^2 + y_2^2 + h^2} \le y_1 + y_2 + h$ for $y_1, y_2 \in (0, +\infty)$, then we have

$$\begin{aligned}\int_0^\infty \int_0^\infty & \frac{\exp\,(c\sqrt{y_1^2 + y_2^2 + h^2})}{r \exp\,(a_1 \operatorname{ch} \rho_1 \alpha)}\, \mathrm{d}y_1\, \mathrm{d}y_2 \\ &\le \int_0^\infty \int_0^\infty \frac{\exp c(y_1 + y_2 + h)}{\exp\,(a_1 \operatorname{ch} \rho_1 \sqrt{(y_1 - x_2)^2 + (y_2 - x_2)^2})} \\ &\qquad \times \frac{\mathrm{d}y_1\, \mathrm{d}y_2}{\sqrt{(y_1 - x_1)^2 + (y_2 - x_2)^2 + (f_1 - x_3)^2}}.\end{aligned}$$

As $\sqrt{y_1^2 + y_2^2 + h^2} \le y_1 + y_2 + h$ for $y_1, y_2 \in (0, +\infty)$, then we have

$$\begin{aligned}\int_0^\infty \int_0^\infty & \frac{\exp c\,(y_1 + y_2 + h)}{\exp\,(a_1 \operatorname{ch} \rho_1 \sqrt{(y_1 - x_2)^2 + (y_2 - x_2)^2})} \\ &\times \frac{\mathrm{d}y_1\, \mathrm{d}y_2}{\sqrt{(y_1 - x_1)^2 + (y_2 - x_2)^2 + (f_1 - x_3)^2}} \\ &= \exp c\,(x_1 + x_2 + h) \int_{-x_1}^\infty \int_{-x_2}^\infty \frac{\exp c(t_1 + t_2)\, \mathrm{d}t_1\, \mathrm{d}t_2}{\sqrt{t_1^2 + t_2^2}\, \exp\,(a_1 \operatorname{ch} \rho_1 \sqrt{t_1^2 + t_2^2})} \\ &\le C_5 \exp\,(c|x|) \int_{-\infty}^\infty \int_{-\infty}^\infty \frac{\exp c(t_1 + t_2)\, \mathrm{d}t_1\, \mathrm{d}t_2}{\sqrt{t_1^2 + t_2^2} \exp\,(a_1 \operatorname{ch} \rho_1 \sqrt{t_1^2 + t_2^2})} \\ &\le C_6 \exp\,(c|x|).\end{aligned} \tag{3.3}$$

This inequality holds since the integral

$$\int_{-\infty}^\infty \int_{-\infty}^\infty \frac{\exp c(t_1 + t_2)\, \mathrm{d}t_1\, \mathrm{d}t_2}{\sqrt{t_1^2 + t_2^2} \exp\,(a_1 \operatorname{ch} \rho_1 \sqrt{t_1^2 + t_2^2})}$$

is converging.

Analogously we obtain the estimate for the rest integrals

$$|I_{1j}| \le C_{j+6} \exp\,(c|x|), \quad j = 2,3,4. \tag{3.4}$$

Taking into account the inequalities (3.3), (3.4), we obtain

$$\left| \int_{\Gamma_1} \Phi(y,x) \frac{\partial U}{\partial n}\, \mathrm{d}s \right| \le C_{11} \exp\,(c|x|). \tag{3.5}$$

The estimate of our integral along the component can be obtained analogously. Thus, inequalities (3.1) and (3.5) yield the estimate

$$|U(x)| \leq C_{12} \exp(c|x|).$$

Suppose that $c = \text{const}$, such that

$$\lim_{R\to\infty} \frac{\max_{|x|=R} |U(x)|}{\exp(c_1/h)|x|} = 0.$$

Now, using Theorem 6.1 [2], we obtain $U(x) \leq 0$. Repeating our considerations to the function $-U(x)$, we obtain that $-U(x) \leq 0$. Hence it follows that $U(x) \equiv 0$, where $x \in \mathcal{D}$. The theorem is proved.

We thank Sh. Ya. Yarmukhamedov for the fruitful discussion.

REFERENCES

1. Z. R. Ashurova, Theorems of Fragmen—Lindelef type for harmonic functions of many variables. *Dokl. Akad. Nauk Uz. SSR* (1990) **5**, 6–8.

2. E. M. Landis, *Elliptic and Parabolic Equations of the Second Order.* Nauka, Moscow, 1971 (in Russian).

3. Sh. Ya. Yarmukhamedov, The Green formula for an infinite domain and its applications. *Izv. Akad. Nauk Uzb. SSR, Ser. Fiz. Mat.* (1981) **5**, 36–42.

Ill-Posed and Non-Classical Problems of Mathematical Physics and Analysis, Samarkand, 2000, pp. 191–197
M. M. Lavrent'ev and S. I. Kabanikhin (Eds)

Recovery of a function set by integrals along a curve family in the plane

A. H. BEGMATOV* and Z. H. OCHILOV*

Abstract — This work deals with the problem of integral geometry [1, Chapter 6] of Volterra type. The integrals are given along the set of parabolas with the special weight function. The stability estimates for solution are obtained in Sobolev's spaces which shows the weak ill-posedness of the problem. Also, the inversion formula is obtained.

Romanov [2] had obtained the uniqueness and stability theorems sufficiently general when the manifolds were paraboloids. The weight functions and manifolds in those cases were invariant relatively the group of all the motions along a fixed hyperplane. The weak ill-posed problems of integral geometry of Volterra type with the weight functions with the singularity were investigated in [3, 4]. The uniqueness theorems, stability estimates, and inversion formulas for weak ill-posed problems of integral geometry when the integration goes along the special curves and surfaces with singularities in vertices are obtained in [5, 6].

Let $\{\mathcal{P}(x,y)\}$ be a set of curves which fill smoothly the half-plane

$$\mathbb{R}^2_+ = \{(x,y):\ x\in\mathbb{R}^1,\ \ y\geq 0\}.$$

Suppose that these curves are parametrized by means of their vertices (x,y) and an arbitrary curve $\mathcal{P}(x,y)$ of this set is defined by the relation

$$\mathcal{P}(x,y) = \{(\xi,\eta):\ y-\eta=(x-\xi)^2,\ \ 0\leq\eta\leq y\}.$$

*This work was partially supported by Uzbek Foundation for Basic Research (grant No 2/01).

Problem 1. *Find the function $u(x,y)$ if for all $(x,y) \in \mathbb{R}^2_+$ the following integrals along the curves $\mathcal{P}(x,y)$ are known:*

$$\int_{\mathcal{P}(x,y)} g(x-\xi)u(\xi, y-(x-\xi)^2)\,\mathrm{d}\xi = f(x,y), \tag{1}$$

$$g(x-\xi) = \begin{cases} 1, & 0 \le x-\xi \le \sqrt{y} \\ 0, & -\sqrt{y} \le x-\xi < 0. \end{cases}$$

Below we shall denote by $\hat{f}(\lambda, y)$ the Fourier transform of the function $f(x,y)$ with respect to the variable x.

Denote by U the class of functions $u(x,y)$ whose partial derivatives up to the second order are continuous and finite with the carrier from $\mathbb{R}^2_+$:

$$\operatorname{supp} u \subset D = \{(x,y):\ -a < x < a,\ 0 < a < \infty,\ 0 < y < l,\ l < \infty\}.$$

Define the right-hand side of equation (1) for $y < 0$. Introduce the function

$$f^*(x,y) = \begin{cases} f(x,y), & y \ge 0 \\ 0, & y < 0. \end{cases}$$

From the setting of the problem and the condition imposed on the function u it follows that we can apply the Fourier transformation to the function $f^*(x,y)$ with respect to y.

Consider now the Fourier transform of the function $f^*(\lambda, y)$ with respect to y. Taking into account that $f^*(x,y)$ vanishes for $y < 0$, we obtain

$$\varphi(\lambda,\mu) = \int_{-\infty}^{\infty} e^{i\mu y} f^*(\lambda, y)\,\mathrm{d}y = \int_0^{\infty} e^{i\mu y} f(\lambda, y)\,\mathrm{d}y.$$

Thus, in the case when we define the function $f(x,y)$ by zero for $y < 0$, we can apply the Fourier transformation both sides of equation (1).

Introduce the functions

$$I(\lambda,\mu) = 2\int_0^{\infty} \frac{e^{i(\mu\tau + \lambda\sqrt{\tau})}}{\sqrt{\tau}}\,\mathrm{d}\tau,$$

$$I_1(\lambda,\mu) = \int_{-\infty}^{\infty} e^{-i\mu y} \frac{\mathrm{d}\mu}{(1+\mu^2) I(\lambda,\mu)}, \qquad I_2 = \int_{-\infty}^{+\infty} e^{-i\lambda x} I_1(\lambda, y)\,\mathrm{d}\lambda.$$

Theorem 1. *Let the function $f(x,y)$ be known for all $y \ge 0$. Then the solution of Problem 1 in the class U is unique, the following representation holds*

$$u(x,y) = \int_{-\infty}^{\infty}\int_{-\infty}^{\infty} I_2(x-\xi, y-\eta)\Big(E - \frac{\partial^2}{\partial\eta^2}\Big) f(\xi,\eta)\,\mathrm{d}\xi\,\mathrm{d}\eta,$$

and the inequality

$$\|u\|_{L_2} \leq C\|f\|_{W_2^{0,1}(R_+^2)}$$

holds, where C is a certain constant and E is the unit operator.

Proof. Apply the Fourier transformation to both sides of equation (1). Denoting $h = \sqrt{y-\eta}$, we can write

$$\int_{-\infty}^{+\infty} e^{i\lambda h} \int_0^y u(x-h,\eta)\frac{\mathrm{d}\eta}{h}\,\mathrm{d}x = \int_0^y \hat{u}(\lambda,\eta)\frac{e^{i\lambda h}}{h}\,\mathrm{d}\eta = \hat{f}(\lambda,y). \qquad (2)$$

We apply to (2) the single Fourier transformation with respect to y:

$$\int_0^\infty e^{i\mu y} \int_0^y \hat{u}(\lambda,\eta)\frac{e^{i\lambda h}}{h}\,\mathrm{d}\eta\,\mathrm{d}y$$
$$= \int_0^\infty \hat{u}(\lambda,\eta) \int_\eta^\infty e^{i\mu y}\frac{e^{i\lambda h}}{h}\,\mathrm{d}\eta\,\mathrm{d}y = \int_0^\infty e^{i\mu y}\hat{f}(\lambda,y)\,\mathrm{d}y.$$

Making the change $\tau = y - \eta$, we obtain

$$v(\lambda,\mu)I(\lambda,\mu) = \varphi(\lambda,\mu), \qquad (3)$$

where

$$v(\lambda,\mu) = \int_0^\infty \hat{u}(\lambda,\mu)e^{i\mu\eta}\,\mathrm{d}\eta, \qquad \varphi(\lambda,\mu) = \int_0^\infty e^{i\mu y}\hat{f}(\lambda,y)\,\mathrm{d}y,$$

$$I(\lambda,\mu) = 2\int_0^\infty \frac{e^{i(\mu\tau+\lambda\sqrt{\tau})}}{\sqrt{\tau}}\,\mathrm{d}\tau. \qquad (4)$$

We show now that integral (4) converges uniformly relatively the parameters λ and μ. Without loss of generality we may consider that the parameter μ is positive.

We see that

$$I = \int_0^\infty e^{i(\mu t^2+\lambda t)}\,\mathrm{d}t = \int_0^\infty \cos(\mu t^2+\lambda t)\,\mathrm{d}t + i\int_0^\infty \sin(\mu t^2+\lambda t)\,\mathrm{d}t.$$

Taking into account the formulas ([7], p. 411)

$$\int_0^\infty \sin(ax^2+2bx)\,\mathrm{d}x = \frac{1}{2}\Big(\cos\frac{b^2}{a} - \sin\frac{b^2}{a}\Big)\sqrt{\frac{\pi}{2a}}, \quad a>0,$$

$$\int_0^\infty \sin(ax^2+2bx)\,\mathrm{d}x = \frac{1}{2}\Big(\cos\frac{b^2}{a} + \sin\frac{b^2}{a}\Big)\sqrt{\frac{\pi}{2a}}, \quad a>0,$$

we obtain

$$I = \frac{1}{2}\sqrt{\frac{\pi}{2\mu}}\Big[\Big(\cos\frac{\lambda^2}{4\mu} + \sin\frac{\lambda^2}{4\mu}\Big) + i\Big(\cos\frac{\lambda^2}{4\mu} - \sin\frac{\lambda^2}{4\mu}\Big)\Big].$$

Estimate the modulus of the function $I(\lambda, \mu)$

$$|I| = \frac{1}{2}\sqrt{\frac{\pi}{2\mu}}\sqrt{\Big(\cos\frac{\lambda^2}{4\mu} + \sin\frac{\lambda^2}{4\mu}\Big)^2 + \Big(\cos\frac{\lambda^2}{4\mu} - \sin\frac{\lambda^2}{4\mu}\Big)^2}.$$

Thus, we have

$$|I| = \sqrt{\pi/\mu}\,/2. \tag{5}$$

Taking into account (5), we obtain from equation (3)

$$v(\lambda, \mu) = \varphi(\lambda, \mu)/I(\lambda, \mu). \tag{6}$$

Divide and multiply the right-hand side of (6) by $q + \mu^2$:

$$v(\lambda, \mu) = (1 + \mu^2)\varphi(\lambda, \mu)/[(1 + \mu^2)I(\lambda, \mu)]. \tag{7}$$

The conditions imposed on the function $u(x, y)$ and (7) yield that the function $\varphi(\,\cdot\,, \mu)$ belongs to L_2 with respect to the argument μ. Equalities (5) and (6) yield that the function $(1 + \mu^2)^{-1}I(\lambda, \mu)^{-1}$ is the Fourier transform of the function $I(\lambda, \mu)$ with respect to y. Apply now the inverse Fourier transformation to equation (7) with respect to μ. Taking into account the inversion theorem, the convolution theorem, and the property of differentiation of Fourier transform, we obtain

$$\hat{u}(\lambda, y) = \int_{-\infty}^{\infty} I_1(\lambda, y - \eta)\Big(E - \frac{\partial^2}{\partial\eta^2}\Big)\hat{f}(\lambda, \eta)\,\mathrm{d}\eta, \tag{8}$$

$$I_1(\lambda, y) = \int_{-\infty}^{+\infty} e^{-i\mu y}\frac{\mathrm{d}\mu}{(1 + \mu^2)I(\lambda, \mu)}.$$

From the restrictions imposed on the function $u(x, y)$ and (7) it follows that the function $f(\lambda, \cdot)$ belongs to L_2 with respect to λ. Applying to equation (8) the inverse Fourier transformation with respect to λ, we obtain the representation of the solution of Problem 1

$$u(x, y) = \int_{-\infty}^{\infty}\int_{-\infty}^{\infty} I_2(x - \xi, y - \eta)\Big(E - \frac{\partial^2}{\partial\eta^2}\Big)f(\xi, \eta)\,\mathrm{d}\xi\,\mathrm{d}\eta,$$

where

$$I_2 = \int_{-\infty}^{\infty} e^{-i\lambda x} I_1(\lambda, y)\, \mathrm{d}\lambda.$$

Relation (5) and formula (7) yield

$$|v(\lambda, \mu)| \leq (1 + |\mu|^2) \cdot |\varphi(\lambda, \mu)|.$$

Hence the estimate follows

$$\int_{-\infty}^{\infty} \int_{-\infty}^{\infty} |v|^2 \, \mathrm{d}\lambda \, \mathrm{d}\mu \leq \frac{1}{\sqrt{\pi}} \int_{-\infty}^{\infty} (1 + |\mu|)^2 |\varphi|^2 \, \mathrm{d}\lambda \, \mathrm{d}\mu. \tag{9}$$

Use now the property of differentiation of the Fourier transform, the triangle inequality for norms, and (8), (9). As a result we obtain

$$\|u\|_{L_2} \leq C \|f\|_{W_2^{0,1}(\mathbb{R}_+^2)}.$$

The theorem is proved. □

We investigate now the problem of integral geometry with perturbation.

Denote by $S(x, y)$ the part of $\mathbb{R}_+^2$ bounded by the curve $\mathcal{P}(x, y)$ and the axis $y = 0$; let Ω_δ be the strip $\{(x, y) : \ x \in \mathbb{R}', \ \ y \in [0, \delta]\}$ where δ is sufficiently small.

Problem 2. *Define the function $u(x, y)$ if for all $(x, y) \in \mathbb{R}_+^2$ we know the sums of the integrals along the curves $\mathcal{P}(x, y)$ and the domains $S(x, y)$:*

$$\int_0^y u(x - h, \eta) \frac{\mathrm{d}\eta}{h} + \int_0^y \int_{x-h}^{x+h} k(x, y, \xi, \eta) u(\xi, \eta) \, \mathrm{d}\xi \, \mathrm{d}\eta = \mathcal{F}(x, y), \tag{10}$$

where h is defined above.

Equation (10) corresponds to the problem of integral geometry with the perturbation. The second term in the left-hand side in the integral with the weight k along the interiors of the parabolas.

Theorem 2. *Suppose that the function $\mathcal{F}(x, y)$ is known for all $y \geq 0$, and the weight function $k \in C_0^2(\Omega_\delta \times \Omega_\delta)$ together with its derivatives up to the second order inclusively vanishes in the parabolas $\mathcal{P}(x, y)$.*

Then the solution of Problem 2 is unique in the class of twice continuously differentiable functions and the inequality

$$\|u\|_{L_2} \leq C_1 \|\mathcal{F}\|_{W_2^{1,0}(\mathbb{R}_+^2)}$$

holds, where C_1 is a certain constant.

Proof. Consider the function

$$f_0(x,y) = \mathcal{F}(x,y) - f(x,y),$$

i.e., the second term from the left hand side of equation (10):

$$\int_0^y \int_{x-h}^{x+h} k(x,y,\xi,\eta)u(\xi,\eta)\,d\xi\,d\eta = f_0(x,y). \tag{11}$$

From the conditions imposed on the functions u and k it follows that the function f_0 will have continuous derivatives up to the second order inclusively.

Taking into account that the function k with its derivatives vanishes in the curves $\mathcal{P}(x,y)$, we have

$$\frac{\partial f_0(x,y)}{\partial x} = \int_0^y \int_{x-h}^{x+h} k_x(x,y,\xi,\eta)u(\xi,\eta)\,d\xi\,d\eta, \tag{12}$$

$$\frac{\partial^2 f_0(x,y)}{\partial x^2} = \int_0^y \int_{x-h}^{x+h} k_{xx}(x,y,\xi,\eta)u(\xi,\eta)\,d\xi\,d\eta. \tag{13}$$

Taking into account the restrictions imposed on the weight function $k(\cdot)$ and the expressions for the derivatives of the function $f_0(\cdot)$, we obtain the estimate

$$\|f_0(x,y)\|_{W_2^{1,2}(\mathbb{R}^2_+)} \le \varepsilon \|u(x,y)\|_{L_2}, \quad 0 < \varepsilon < 1. \tag{14}$$

The integral operators standing in the left-hand sides of the equations (1) and (11) we denote by A_0 and A_1. Equations (1) and (10) in this case will become as follows:

$$A_0 u = f,$$

$$A_0 u + A_1 u = \mathcal{F}. \tag{15}$$

As it follows from [8], the operator A_0 has the left inverse A_0^{-1}. Acting by A_0^{-1} on the both sides of (15) we obtain

$$u + A_0^{-1} A_1 u = A_0^{-1}\mathcal{F}. \tag{16}$$

From the estimates of Theorem 1 it follows that the operator $A_0^{-1}A_1$ from (16) is continuous. Therefore, the operator $A_0^{-1}A_1$ satisfies the estimate

$$\|A_0^{-1}A_1\| \le \varepsilon < 1. \tag{17}$$

From the contraction mapping principle being applied for the operator from the right-hand side (10) follows the uniqueness of solution of equation (10). Thus, inequalities (14), (17) and Theorem 1 yield the estimate

$$\|u\|_{L_2} \leq C_1 \|\mathcal{F}\|_{W_2^{0,1}(\mathbb{R}_+^2)},$$

where C_1 is a certain constant. □

REFERENCES

1. M. M. Lavrent'ev, V. G. Romanov, and S. P. Shishatskii, *Ill-Posed Problems of Mathematical Physics and Analysis.* Nauka, Moscow, 1980 (in Russian).

2. V. G. Romanov, *Some Inverse Problems for the Equations of Hyperbolic Type.* Nauka, Novosibirsk, 1972 (in Russian).

3. A. H. Begmatov, *Sib. Math. J.* (1995) **36**, No. 2, 243–247.

4. A. H. Begmatov, *J. Inv. Ill-Posed Problems.* (1995) **3**, No. 3, 231–235.

5. A. H. Begmatov, *Sib. Math. J.* (1997) **38**, No. 4, 723–737.

6. A. H. Begmatov, *Dokl. Ros. Akad. Nauk* (1998) **358**, No. 2, 151–153 (in Russian).

7. I. S. Gradshtein and I. M. Ryzhik. *Tables of Integrals, Summs, Series and Product.* Fizmatgiz, Moscow, 1962 (in Russian).

8. S. G. Krein (Ed), *Functional Analysis.* Reference Mathematical Bibliography. Nauka, Moscow (in Russian).

Ill-Posed and Non-Classical Problems of Mathematical Physics and Analysis, Samarkand, 2000, pp. 199–205
M. M. Lavrent'ev and S. I. Kabanikhin (Eds)

Uniqueness of extension of solutions of differential equations of the second order

A. HAIDAROV and D. SHODIEV

Abstract — For a linear differential equation of the second order the Cauchy problem with the data in the cylinder surface of the time type is considered.

Some important problems of physics and geophysics are connected with the uniqueness problem for the Cauchy problem for differential equations [1]. The first uniqueness theorem for extension of solutions of equations with nonanalytic coefficients belongs to T. Carleman [2]. The method was based on establishment of the domain weight estimates dependent on the domain and differential operators. This method was improved by F. John [3], M. M. Lavrant'ev [4, 5], L. Nirenderg [6], L. Hörmander [7] et al. For the nonanalytic hyperbolic equations with the wave operator in the principal part and the carrier of the form $(-T, T) \times S$, the domain of dependence was found in [8] if S is the sphere of the radius T. The uniqueness of extension into a certain neighborhood Γ was established in the case when S is strictly convex in [9]. When $\Gamma = (-T, T) \times S$ and S is a hyperplane in E^n of the class C^2, for the research of the Cauchy problem for hyperbolic equation with the wave operator in the principal part in [10] the general theorems of L. Hörmander [7, Chapter 1] were used. Also, in [10], for the case of the local problem (with the data in the part of boundary of the cylinder domain) the case when the wave coefficient c is equal to 1 was considered. In [11] the special case of the problem considered in [10] with the variable wave coefficient was developed. The local Cauchy problem for a certain class of hyperbolic equations was considered in [12], and the estimates and the uniqueness theorems were established. In our paper, the problem of unique extension of the solution of one class of differential second order equations from a part of the boundary was considered. The stability estimate in a neighborhood of the boundary Γ was obtained for a weaker suppositions relatively the coefficients of the equation then in [11, 12].

Introduce the notations: $x = (x_0, \dots, x_n)$ is a vector from $\mathbb{R}^{n+1}$; $t = x_0$, $x' = (0, x_1, \dots, x_{n-1}, 0)$;

$$\langle x, y\rangle = x_0 y_0 + \dots x_n y_n.$$

For an arbitrary set B, we denote by $B_\sigma = B \cap \{\psi > \sigma\}$, where ψ is the function defined below. Let N be an exterior normal to the boundary of the domain, $\alpha = (\alpha_0, \dots, \alpha_n)$; $D^\alpha = D_0^{\alpha_0} \cdots D_n^{\alpha_n}$, where $D_j = -i\partial/\partial x_j$, $\|\cdot\|_{(k)}(Q)$ is the norm in the Sobolev space $W_2^k(Q)$; $|\cdot|(Q)$ is the norm in the space $C^k(Q)$ of the functions k times continuously differentiable. Let $\overset{\circ}{W}{}_2^k$ be the closure of $C_0^\infty(Q)$ with respect to the norm $\|\cdot\|_{(k)}(Q)$. Suppose also that Ω is a domain from $\mathbb{R}^n$ whose boundary consists from a part of the hyperplane $x_n = d$ and a component S from the class C^2 which lies in the layer $-d < x_n \le 0$. Denote

$$Q = (-T, T) \times \Omega, \qquad \Gamma = (-T, T) \times S,$$

where T, d are positive constants and $R_0 = \sup_\Omega |x'|$.

We consider the problem of uniqueness and stability of solution of the following Cauchy problem

$$Au = f \quad \text{in} \quad Q_0, \tag{1}$$

$$u = g_0, \quad \frac{\partial u}{\partial N} = g_1 \quad \text{in} \quad \Gamma_0, \quad u \in W_2^2(Q_0), \tag{2}$$

where

$$A = \frac{\partial^2}{\partial t^2} - \sum_{j,k=1}^{n} a^{jk}(x) \frac{\partial^2}{\partial x_j \partial x_k} + \sum_{j=0}^{n} a^j(x) \frac{\partial}{\partial x_j} + a.$$

The coefficients of A satisfy the following conditions

$$a^{jk} \in C^1(Q), \qquad a^{nn} \ge \varepsilon_0, \qquad a^{jn} = 0 \quad \text{for} \quad j \le n-1,$$
$$a^{jk} = a^{kj}, \qquad a^j, a \in L_\infty(Q). \tag{3}$$

Define the function ψ as follows:

$$\psi(x) = \exp\left[\lambda(x_n + d(1 - x_0^2 T^2)) + 2^{-1}\delta(|x'|^2 - R_0^2)\right] - 1, \tag{4}$$

where the parameters λ and δ are chosen below. Denote also the operators ∇' and A_{x_k}:

$$\nabla' = \left(0, \frac{\partial}{\partial x_1}, \dots, \frac{\partial}{\partial x_n}\right), \qquad A_{x_k} = \frac{\partial}{\partial x_k} A.$$

Theorem 1. *Let*

$$\langle A\xi', \nabla' \langle Ax', \xi' \rangle\rangle \geq \varepsilon_1 |\xi|^2, \qquad \langle A_{x_n} \xi', \xi' \rangle \leq 0,$$
$$\langle A\xi', \nabla' \langle Ax', \xi' \rangle\rangle + 2(a^{nn})^{-1} \langle \nabla' a^{nn}, Ax' \rangle \langle Ax', \xi' \rangle \leq 2^{-1} \varepsilon_1 |\xi'|^2, \tag{5}$$

for all $\xi \in \mathbb{R}^{n+1}$, $x \in Q$.

Then there exists such a number M *dependent only on* $|a^{jk}|^1(Q)$, *so that for*

$$|A_{x_0} x'| < \varepsilon_1 \varepsilon_0 M^{-1}, \quad dT^{-1} < \delta \varepsilon_1 \varepsilon_0^2 M^{-1}, \quad \delta < \varepsilon_1 \varepsilon_0^2 M R_0^2, \quad M\varepsilon^{-2} < \lambda, \tag{6}$$

the solution of problem (1), (2) *satisfies the inequality*

$$\|u\|_{(1)}(Q_\varepsilon) \leq K(\varepsilon) \|u\|_{(2)}^{1} [\|f\|_{(0)}(Q) + \|g_1\|_{(1)}(\Gamma_0) + \|g_0\|_{(2)}(\Gamma_0)]^{\mu}, \tag{7}$$

where K *is a certain constant dependent on* ε *only and* $\mu = \varepsilon(2 \exp \lambda d - 2 - \varepsilon)^{-1}$.

This theorem is established with the use of the following estimate of the Karleman type.

Theorem 2. *Under the conditions of Theorem 1, there exists a constant* K, *so that*

$$\sum_{|\alpha|<2} \tau^{3-|\alpha|} \int_Q |D^\alpha u|^2 e^{2\tau\psi}\, dx \leq K \int_Q |Au|^2 e^{2\tau\psi}\, dx,$$
$$\tau \geq K \qquad \text{for all} \quad u \in \mathring{W}_2^2(Q_0). \tag{8}$$

To prove Theorem 2 we will use the conditional Theorem 8.4.3 of Hörmander [7]. In order to simplify the calculations we introduce the following change of variables by the formulas:

$$y_0 = x_0, \qquad \dots, \qquad y_{n-1} = x_{n-1},$$
$$y_n = x_n + d(1 - T^{-2} x_0^2) + 2^{-1} \delta (|x'|^2 - R_0^2). \tag{9}$$

Denote by P the characteristic polynomial of the principal part of the operator A in the variables y. The functions a^{jk} we denote as b^{jk} and ψ we denote by φ. As a result we obtain

$$P(y,\zeta) = -\zeta_0^2 + \sum_{j,k=1} b^{jk}\zeta_i\zeta_k + B_0\zeta_0\zeta_n + 2\delta\langle By'\zeta\rangle\zeta_n + F\zeta_n^2,$$
$$\varphi = e^{\lambda y_n} - 1, \qquad \Delta\varphi = (0,0,\dots,\lambda e^{\lambda y_n}),$$
$$\zeta = (\xi_0,\xi_1,\dots,\xi_{n-1},\xi_n + i\gamma), \qquad \gamma = \tau\lambda e^{\lambda y_n}, \tag{10}$$

where

$$B_0 = 4dT^{-2}, \qquad F = b^{nn} - 4d^2T^{-4}y_0^2 + \delta^2\langle By', y'\rangle, \tag{11}$$

and B is a linear operator in $\mathbb{R}^n$ with a matrix (b^{jk}). Set

$$P^{(j)}(y;\zeta) = \frac{\partial P(y;\zeta)}{\partial\zeta_j}, \qquad P_{(j)}(y;\zeta) = \frac{\partial P(y;\zeta)}{\partial y_j}. \tag{12}$$

Introduce the following form quadratic with respect to ξ

$$H(y,\zeta) = \sum_{j,k=0}^{n} \frac{\partial^2\varphi}{\partial y_j\partial y_k} \cdot P^{(j)}(y;\zeta)\cdot\overline{P^{(k)}(y;\zeta)} + \tau^{-1}\operatorname{Im}\sum_{k=0}^{n} P_{(n)}(y;\zeta)\cdot\overline{P^{(k)}(y;\zeta)}, \tag{13}$$

where $\zeta = \xi + i\tau\nabla\varphi(y)$.

Let G_0 be the image of Q_0 under the mapping (9).

Lemma 1. *If the vector $\zeta \neq 0$ satisfies the characteristic equation $P(y;\zeta) = 0$ and conditions (5) and (6) of Theorem 1 hold, then the form $H(y;\zeta)$ is positive determine for all $y \in G_0$.*

Proof. Taking into account formulas (10)–(12), the polynomials $P^{(j)}$ and $P_{(j)}$, $j = \overline{0,n}$ are defined by the formulas

$$P^{(0)} = -2\xi_0 + B_0\zeta_n, \quad P^{(j)} = 2(B\xi')_j + 2\delta(By')_j\zeta_n, \quad 1 \le j \le n-1,$$
$$P^{(n)} = B_0\xi_0 + 2\delta\langle By',\xi'\rangle + 2F\zeta_n, \tag{14}$$
$$P_{(j)} = \langle B_{y_j}\xi',\xi'\rangle + B_{0y_j}\xi_0\zeta_n + 2\delta\langle (By')_{y_j},\xi'\rangle\zeta_n + 2F\zeta_n^2. \tag{15}$$

Separating the real part of the equation $P(y;\zeta) = 0$ and taking into account that $\tau^{-1}\operatorname{Im}P(y;\zeta) = 0$ and (10), we obtain

$$-\xi_0^2 + \langle B\xi',\xi'\rangle + (B\xi_0 + 2\delta\langle By',\xi'\rangle)\xi_n + F(\xi_n^2 + \gamma^2) = 0,$$
$$B_0\xi_0 + 2\delta\langle By',\xi'\rangle + 2F\xi_n^2 = 0. \tag{16}$$

Taking into account (16), rewrite the last equality as follows:

$$\langle B\xi', \xi'\rangle = \xi_0^2 + F\xi_n^2 + F\gamma^2. \tag{17}$$

Changing now T and δ so that

$$4dT^{-1} < \sqrt{\varepsilon_0}, \qquad \delta^2|\langle By', y'\rangle| < \varepsilon_0/4$$

and taking into account (3) and (11), we obtain

$$F > \varepsilon_0/2. \tag{18}$$

This and (16), (17), (18) yield the inequalities

$$|\xi_0| \le C_1|\xi'|, \qquad |\xi_n| \le L|\xi'|, \tag{19}$$

where $C_1 = \sqrt{nM_0}$, $L = \varepsilon^{-1}(\sqrt{nM_0}\,|B_0| + 2nM_1|y'|\delta)$ and the constants M_l $(l = 0, 1)$ are defined as follows:

$$M_l = \sup_Q |\nabla^l a^{jk}|.$$

From (14) and (15) by immediate computation we obtain

$$\begin{aligned}\tau^{-1}\operatorname{Im} P_{(0)} \cdot \bar{P}^{(0)} = \lambda e^{\lambda y_n}[-2B_0 y_0 \xi_0^2 - 4\delta\langle B_{y_0} y', \xi'\rangle\xi_0 - B_0\langle B_{y_0}\xi', \xi'\rangle \\ - 4F_{y_0}\xi_0\xi_n + B_0 F_{y_0}(\xi_n^2 + \gamma^2)],\end{aligned}$$

$$\begin{aligned}\tau^{-1}\operatorname{Im}\sum_{j=1}^{n-1} P_{(j)} \cdot \bar{P}^{(j)} = \lambda e^{\lambda y_n}[-2\delta\langle By', \nabla'\langle B\xi', \xi'\rangle\rangle + 4\delta\langle B\xi', \nabla'\langle By', \xi'\rangle\rangle \\ + 4\langle\nabla' F, B\xi'\rangle\xi_n + 2\delta\langle By'\nabla' F\rangle(\xi_n^2 + \gamma^2)],\end{aligned}$$

$$\begin{aligned}\tau^{-1}\operatorname{Im} P_{(n)} \cdot \bar{P}^{(n)} = \lambda e^{\lambda y_n}[2\delta B_0\langle B_{y_n} y', \xi'\rangle\xi_0 - 2F\langle B_{y_n}\xi', \xi'\rangle \\ +4\delta^2\langle B_{y_n} y', \xi'\rangle\cdot\langle By', \xi'\rangle + (4\delta F_{y_n}\langle By', \xi'\rangle + 2B_0 F_{y_n}\xi_0)\xi_n + 2FF_{y_n}(\xi_2^2 + \gamma^2)].\end{aligned}$$

Using the definition (13) of the form H and the equality $|P^{(n)}|^2 = 4F^2\gamma$, which follows (16) and inequalities (19), we deduce the estimate

$$\lambda^{-1}e^{-\lambda y_n} H(y;\zeta) \ge \lambda\varepsilon_0^2\gamma^2/2 + \delta\varepsilon_1|\xi'|^2/2.$$

similar to those obtained in [12].

Thus, $H(y;\zeta) \ge 0$ in the characteristic vector ζ, where $\gamma = 0$, $|\xi'| = 0$ if $H(y;\zeta) = 0$. Therefore, since $\zeta_n = 0$, we have $\zeta = 0$. □

Lemma 1 yields the conditions of Theorem 8.4.3 from [7] for the operator A with the characteristic symbol $P(y;\zeta)$ in (10). Therefore, the estimate (8) holds with $G = G_0$. Change the variables by formulas (9) in integrals (8). Taking into account that Jacobian of this change is equal to unity and that estimate (8) admits the addition of junior terms to $P(y;\zeta)$, we see that Theorem 2 holds.

Theorem 1 can be proved from Theorem 2 with the use of the well-known scheme [1, p. 96]. In the case when the principal part of A is the wave operator, the proof of Theorem 1 can be found in [11]. In our case the proof of Theorem 1 is similar. The estimate (7) of Theorem 1 yields the uniqueness of solution of the Cauchy problem (1), (2) in Q_0. The character of dependence of the constant K on ε in (7) in the case of the operator A with the wave operator as the principal part is considered in [7]. From [7] it follows that such dependence K on ε allows us to prove the uniqueness of the problem in question. Theorem 1 clarifies the geometry of the domain Ω and the form of the operator A for which the problem is conditionally well-posed. Estimate (7) of Theorem 1 is necessary also for research of inverse problems on determination of coefficients of A in cylinder domains. For hyperbolic equations such estimate is applied for research of inverse problems in [13].

REFERENCES

1. M. M. Lavrent'ev, V. G. Romanov, and S. P. Shishatskii, *Ill-Posed Problems of Mathematical Physics and Analysis.* Nauka, Moscow, 1980 (in Russian).

2. T. Carleman, Sur un probleme de unicite pour des sistemes d'equtions aux derivees partitielles deux variables independentes. *Ark Math. Astronom. Fys.* (1939) **17**, 1–9.

3. F. F. John, Continuous dependence on data for solution of partial differential equations with a prescribed bound. *Comm. Pure. Appl. Math.* (1960) **13**, 551–885.

4. M. M. Lavrent'ev, On the Cauchy problem for the Laplace equation. *Izv. Akad. Nauk SSSR. Ser. Mat.* (1956) **20**, 819–842 (in Russian).

5. M. M. Lavrent'ev, On the Cauchy problem for linear elliptic equations of the second order. *Izv. Akad. Nauk SSSR. Ser. Mat.* (1956) **20**, 819–842 (in Russian).

6. L. Nirenberg, Lectures on linear differential equations with partial derivatives. *Uspekhi Mat. Nauk* (1975) **30**, No. 4, 147–204 (in Russian).

7. L. Hörmander, *Linear Differential Operators with Partial Derivatives.* Mir, Moscow, 1965 (in Russian).

8. A. C. Murray and M. H. Protter, Asymptotic behavior and Cauchy problem for ultrahyperbolic equations. *Indiana Univ. Math. J.* (1974) **24**, No. 2, 115–130.

9. S. P. Shishatskii, A priori estimates in the problem on extension of the wave field from the cylinder time-like surface. *Dokl. Akad. Nauk SSSR* (1973) **213**, No. 1, 49, 50.

10. V. M. Isakov, On uniqueness of extension of solutions of hyperbolic equations. *Mat. Zametki* (1982) **32**, No. 1, 75–81 (in Russian).

11. A. Khaidarov, *Carleman Estimates and Uniqueness Theorems for the Coefficient c of the Operator* $cu_{tt} - \Delta u$. Preprint **475**. Calcul. Center, Siberian Branch of the USSR Acad. Sci., Novosibirsk, 1983 (in Russian).

12. A. Khaidarov, Carleman estimates and inverse problems for second order hyperbolic equations. *Mat. Sbornik* (1986) **130**, No. 2, 265–274 (in Russian).

13. A. Khaidarov, Stability estimates in inverse problems for hyperbolic equations. *Mat. Zametki* (1991) **49**, No. 1, 119–144 (in Russian).